LEGO® MINDSTORMS® NXT
THINKING ROBOTS

LEGO® MINDSTORMS® NXT THINKING ROBOTS

build a rubik's cube® solver and
a tic-tac-toe playing robot!

daniele **benedettelli**

no starch
press

Printed in the United States of America

13 12 11 10 09   1 2 3 4 5 6 7 8 9

ISBN-10: 1-59327-216-2
ISBN-13: 978-1-59327-216-6

Publisher: William Pollock
Production Editor: Phil Dangler
Cover and Interior Design: Octopod Studios
Cover Photograph: Francesco Rossi
Developmental Editor: William Pollock
Technical Reviewers: Josh Pollock, Keith Fancher, and Leigh Poehler
Copyeditor: Marilyn Smith
Compositor: Riley Hoffman
Proofreader: Liz Welch

For information on book distributors or translations, please contact No Starch Press, Inc. directly:

No Starch Press, Inc.
555 De Haro Street, Suite 250, San Francisco, CA 94107
phone: 415.863.9900; fax: 415.863.9950; info@nostarch.com; www.nostarch.com

Library of Congress Cataloging-in-Publication Data

Benedettelli, Daniele, 1984-
   LEGO MINDSTORMS NXT thinking robots : build a Rubik's cube solver and a tic-tac-toe playing robot! / Daniele Benedettelli.
      p. cm.
   Includes index.
   ISBN-13: 978-1-59327-216-6
   ISBN-10: 1-59327-216-2
   1. Robots--Design and construction. 2. LEGO toys. I. Title.
   TJ211.B4634 2009
   629.8'92--dc22
                                    2009040818

To my parents and grandparents

brief contents

contents in detail

acknowledgments

First of all, in order to avoid missing anyone, thanks to all who helped with this book. It may not seem like it, but this little book is just the tip of a huge iceberg. In fact, it's the product of three years' work, especially in the case of the Rubik's Cube solver.

Thanks first to my cousin Riccardo, who induced me to try to tackle the Rubik's Cube problem. (I mean seriously solving it, not by swapping stickers as we did as kids!) At the time, I was studying artificial intelligence at university, and the idea occurred to me to create a simple LEGO robot that could solve a cube autonomously and very quickly. Thanks also to Professor Marco Gori for inviting me to the 2007 Festival of Creativity in Florence, where I demonstrated my robot for the first time. Much of the robot's success, at least in Italy, sprang from that festival.

I can't forget to thank Narendra Gaonkar! Together, we discussed titles and names for this book's robots.

Thanks and congratulations to Francesco Rossi, who took the photos of my robots, including the great cover shot.

Thanks to the children who helped me to make sure that the robots were really child-friendly by building and playing with them, and to Josh Pollock for endlessly building and testing the final models. And, of course, a big thank you to everyone at No Starch Press, especially Riley, Alison, and Bill.

For their patience and support, a huge thanks to my family. During the final stages of this project I was almost unbearable, uttering mostly inattentive monosyllables.

And finally, thanks to all of you who have waited a long time for the publication of my LEGO Rubik Utopia project. Your endless requests spurred me on to complete the book that you now hold in your hands.

Play well! (Leg godt!)

introduction

With the help of this book, you will build two robots that think! Hard to believe? Well, imagine a robot that can play tic-tac-toe flawlessly and never lose a single match, or a robot that can solve the brain-teasing Rubik's Cube® in just a few minutes. Whether you have the LEGO® MIND-STORMS® NXT 2.0 set (8547) or are the proud owner of the original LEGO MINDSTORMS NXT retail set (8527), you won't need to worry about your LEGO parts resources. This book contains detailed instructions for building both the TTT Tickler (a fierce tic-tac-toe opponent) and the One-Armed Wonder (a Rubik's Cube solver) with either LEGO MINDSTORMS set (two versions for each robot, depending on your set), with no need for extra LEGO parts or expense. The programs for the robots are available from *http://tr.benedettelli.com/* or *http://www.nostarch.com/*.

who is this book for?

This book is for anyone—novices or advanced users—who wants to tackle the amazing world of robotics. Therefore, I've tried to strike a balance between technical details and simplicity.

Explaining the abstract concepts and complex programming behind these thinking robots would require another book entirely, so to keep things simple and fun I won't delve too deeply into the complex programming behind each robot's artificial intelligence. However, in the appendix, you'll find additional details about the programming, as well as some hints, a bibliography, and web resources for more information about developing your own thinking robots.

what's needed to use this book?

In order to build the robots in this book, you will need the original LEGO MINDSTORMS NXT (8527) or the LEGO MINDSTORMS NXT 2.0 set (8547). (The original set is often referred to as the 1.0 set.) Each set contains all the parts required for the robots. You'll also need a computer, with the minimum requirements described in the LEGO MIND-STORMS NXT user manual included with your set.

The TTT Tickler is computer independent: once you download the program to it, the robot will play on its own. On the other hand, the One-Armed Wonder won't work without a USB connection to a computer. That computer will need to run its control program (downloadable from *http://tr.benedettelli.com/* or *http://www.nostarch.com/*) on a computer running Windows 2000 or later.

The One-Armed Wonder also needs a webcam in order to take the photos of the cube. The robot was designed to hold a Logitech QuickCam Chat, which is a bit old but available through online retailers. I chose the QuickCam Chat because it's spherical, with a hole at the bottom and two holes on the sides, which makes it a perfect fit for LEGO parts. As of this writing, similar webcams are the Logitech Webcam C200, C250, or C500. You can also use the LEGO Logitech Camera, included in the LEGO MINDSTORMS Vision Command 9731 set, released in 2000. Any webcam will do, but you should modify the robot design to fit your hardware.

Also, when choosing a webcam, be sure that it works with your operating system.

Oops, I almost forgot! In order to play tic-tac-toe with your robot, you can use the colored Zamor Spheres included in the NXT 2.0 set. The robot will use the red and yellow ones, and you'll use the blue and green ones. If you're using the NXT 8527 set, you could use some glass marbles that are the same size as the Zamor Spheres: five dark marbles and five light-colored ones.

Both versions of the One-Armed Wonder need a 65mm × 55mm cardboard or plastic slippery sheet. The One-Armed Wonder version 1.0 requires two extra rubber bands the same size as the white LEGO rubber bands. And, of course, both models need a Rubik's Cube to solve! The best cubes for our purposes are the original ones, which are easier to manipulate than some of the knockoffs. You can easily recognize the knockoffs, because their faces are hard to turn, they get stuck easily (you'll risk having your robot get stuck or even shatter the cube), their colors are not attractive, and their color scheme doesn't match the original. The original color scheme is yellow opposite white, red opposite orange, and blue opposite green.

This said, happy building!

1

getting started

Since you're reading this book, I assume that you already know what the NXT is. Therefore, I'll spare you another boring introduction that details what you can do with the LEGO MINDSTORMS system, its possibilities, its features, and its flaws. I'll begin instead with an introduction to the software that you'll use to control the robots in this book. Instead of using the NXT-G software that comes with your NXT system, Windows users will learn how to use Bricx Command Center (BricxCC), and Mac users will learn how to use NeXT Tools and the NBC compiler for the Mac.

You'll build two models in this book: the TTT Tickler (which will play tic-tac-toe with you) and the One-Armed Wonder (a robot that can solve the Rubik's Cube). Both models require pretty complicated programs, which can't be built easily with the NXT-G graphical programming language. Instead, I've chosen to use the powerful textual programming language called NXC (Not eXactly C), which makes developing and maintaining the code much simpler than it would be with NXT-G.

This chapter will show you how to use BricxCC to write, compile, and test NXC programs. BricxCC also includes many utilities for use with the NXT brick, including the following:

* A browser to explore the contents of the NXT's memory
* A sound converter to translate sound files in the WAV file format into the RSO file format for the NXT
* A panel to monitor the NXT's sensors and outputs
* A screen-capture utility

BricxCC was originally written by Mark Overmars, and is now magnificently maintained by John Hansen, the creator of the NXC language. John Hansen also developed the NeXT Tools for Mac, a collection of utilities for the NXT that are the BricxCC counterparts for Mac OS X.

I will not teach you how to program with NXC in this book, but if you would like to learn how to write programs in NXC, I recommend reading my tutorial on the subject, as well as the excellent NXC programming guide by John Hansen. Together, these make a comprehensive manual for learning to program in NXC. You can download the NXC programmer's guide, my tutorial, and sample NXC code from *http://bricxcc.sourceforge.net/nbc/nxcdoc/index.html*. To download the NXC manual directly, visit *http://bricxcc .sourceforge.net/nbc/nxcdoc/NXC_Guide.pdf*. My NXC tutorial is available directly from *http://bricxcc.sourceforge.net/nbc/ nxcdoc/NXC_tutorial.pdf*.

sound and program files

In order to use the robots you'll build in this book, you'll need to copy various files to your NXT brick. You can download these files from *http://tr.benedettelli.com/* or from *http:// www.nostarch.com/*. The archive contains the files listed in Table 1-1.

table 1-1: files provided for download

File	Description
! Attention.rso	Sound file
! Blips 19.rso	Sound file
! Click.rso	Sound file
! Fanfare.rso	Sound file
! Sonar.rso	Sound file
! Startup.rso	Sound file
Crying 02.rso	Sound file
Goodbye.rso	Sound file
Laughing 02.rso	Sound file
Play.rso	Sound file
Try Again.rso	Sound file
ttt.ric	Graphics file for the TTT Tickler
ttt_tickler_1.rxe	TTT Tickler program for firmware 1.05
ttt_calib_1.rxe	TTT Tickler calibration utility for firmware 1.05
ttt_tickler_1.rxe	TTT screen-based program for firmware 1.05
LRU09_1.rxe	One-Armed Wonder program for firmware 1.05
ttt_tickler_2.rxe	TTT Tickler program for firmware 1.28
ttt_calib_2.rxe	TTT Tickler calibration utility for firmware 1.28
ttt_tickler_2.rxe	TTT screen-based program for firmware 1.28
LRU09_2.rxe	One-Armed Wonder program for firmware 1.28

setup for BricxCC or NeXT Tools

In this section, I'll guide you through installing the software you'll use to manage your NXT brick and copy programs to it. You should read this section whether you work under Windows or Mac OS X.

To begin, download the LEGO NXT drivers from *http://mindstorms.lego.com/support/files/*. Look for the **MINDSTORMS NXT Driver v1.02**, and be sure to download the correct version for your operating system (Windows or Mac). If you are running Mac OS X 10.5 or later, download and install the Mac OS 10.5 (Leopard) Firmware Fix from the same updates page. You'll find instructions for installing each file on the website.

If you want to get the best from your NXT, you will need the latest NXT-enhanced firmware (version 1.28 or higher).

To install it, download it from *http://bricxcc.sourceforge.net/lms_arm_nbcnxc.zip*, and then extract this Zip file to your desktop. (You will need the file *lms_arm_nbcnxc_128.rfw* contained in this Zip file when it's time to upgrade your firmware.)

In the following sections, I'll show you how to write, compile, and download your first NXC program to the NXT, using BricxCC or the command-line compiler for Mac. Next, I'll show you how to copy files to the NXT using either BricxCC or NeXT Tools.

quick start guide for Windows users

This section is for Microsoft Windows users only. It applies to Windows 2000 and higher.

To install BricxCC under Windows, first download the latest BricxCC release from *http://bricxcc.sourceforge.net/*. When choosing the file to download from this page, click the link titled **latest version** under the Downloading heading. The direct link to this file should be *http://sourceforge.net/projects/bricxcc/files/bricxcc/*.

Choose the latest release from the list (3.3.7.19 as of this writing) and then double-click the downloaded executable installer and follow the onscreen instructions. Optionally, to update the program to its latest version, download the test release from *http://bricxcc.sourceforge.net/test_release.zip* and extract it directly into the BricxCC installation folder (for example, *C:\Program Files\BricxCC*). BricxCC should be compatible with all 32-bit versions of Microsoft Windows (2000 and higher).

You should have the NXT driver version 1.02 or later installed on your system and your NXT should be turned on and connected to the PC via USB.

To begin, open BricxCC by clicking its icon in the Windows Start menu. When BricxCC starts up, you will see the Find Brick dialog, as shown in Figure 1-1. You will use this dialog to tell BricxCC which kind of brick you're using (BricxCC can manage various programmable bricks produced by LEGO throughout the years) and how the brick is connected.

In order to access your NXT with BricxCC, choose **NXT** from the Brick Type drop-down list, select **Automatic** from the Port list, and leave the Firmware radio button set to

Figure 1-1: The Find Brick dialog appears when you open BricxCC.

Standard, as shown in Figure 1-1. Click **OK**, and BricxCC should find the brick.

If everything goes well, you should not see any error messages, and the toolbar icons should be enabled (as shown in Figure 1-2) as opposed to grayed out (as in Figure 1-3).

Figure 1-2: The BricxCC interface with buttons enabled, when the NXT is connected

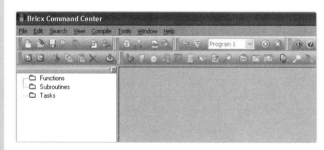

Figure 1-3: The BricxCC interface with buttons disabled, meaning the NXT is not connected

At this point, you should be connected to your brick. Now let's see what you can do in the BricxCC integrated development environment (IDE).

creating NXC programs

To create a new file to contain your NXC program, use one of these methods:

* Press CTRL-N.
* Select **File ▸ New** from the menu.
* Click the new file icon on the toolbar.

    Then double-click the file's title bar to expand it.

NOTE Double-clicking is required only if the MDI mode is enabled. See general preferences.

For example, the following trivial NXC program prints "Hello, World!" on the screen, waits for two seconds, and then ends.

```
task main ()
{
    TextOut (10,LCD_LINE4,"Hello, World!");
    Wait (SEC_2);
}
```

Save the file using one of the following methods:

* Press CTRL-S.
* Select **File ▸ Save**.
* Click the save file icon on the toolbar (the third icon from the left, showing the floppy disk).

You'll be asked for a name and location in which to save the file. Name it *hello.nxc*, save it to your desktop (or any location that you prefer), and click **OK**.

enabling the internal compiler

The latest BricxCC release has a built-in NXC compiler. To enable it, open the Preferences dialog (by clicking the edit preferences icon on the toolbar or choosing **Edit ▸ Preferences** from the menu). Go to the Compiler tab and select the NBC/NXC subtab, then check the **Use internal compiler** box and click **OK**.

To confirm that you are using the correct version of the NBC compiler, return to the NBC/NXC subtab under the Compiler tab in the Preferences dialog and click the **Version** button. You should have NBC compiler version 1.0.1.35 or later.

compiling NXC programs

Now that your file has an NXC extension, you can compile it. To do so, use one of the following methods:

* Press F5.
* Select **Compile ▸ Compile** from the menu.
* Click the compile program icon on the toolbar.

If there are no compilation errors, BricxCC will be silent—it should simply compile the program using the NXC compiler and your brick's connection settings.

NOTE In the Compiler Options tab, check the option NXT 2.0 Compatible Firmware if you are compiling for an NXT running firmware 1.28.

downloading and running NXC programs

Your program is now ready to be downloaded to the NXT. To download and start it, press CTRL-F5 or select **Compile ▸ Download and Run** from the menu. (To download it without running it, press F6, or select **Compile ▸ Download**.) The NXT should beep, start your program, and display the text "Hello, World!"

You can find other commands to start and stop the program in the same Compile menu. If you ever feel lost, just hover the mouse over a toolbar icon to get the pop-up tip, or consult the BricxCC online guide.

using BricxCC tools

BricxCC includes many useful tools to manage your NXT brick. The following are the ones you will need in this book:

Diagnostics This utility tells you all available information about the NXT connected to your computer, including the firmware version, battery voltage, connection type (USB or Bluetooth), its name and Bluetooth address, and the amount of free memory in bytes.

NXT Explorer This tool is a file browser for the NXT flash memory. It allows you to upload, download, and delete files from the NXT, and even defragment the NXT's memory. You can use the NXT Explorer to play a sound file or to start a compiled program (with the *.rxe* extension).

NXT Screen This utility allows you to view and capture the contents of the NXT's LCD screen on your PC. Use it to view an out-of-reach NXT screen or to remotely press an NXT's buttons.

Find Brick/Turn Brick Off/Close Communication These items should be self-explanatory. Find Brick opens the startup panel, where you can choose how to connect the NXT. Turn Brick Off turns off the brick. Close Communication releases the connection and corresponding drivers.

Download Firmware Use this tool to update the NXT firmware.

downloading files to the NXT

Now that you've been introduced to the basics of BricxCC, you're ready to download the files you'll need to control the robots you'll build in this book to your NXT.

First, download the archive containing the files for this book from *http://tr.benedettelli.com/* or from *http://www.nostarch.com/*, then unzip the downloaded archive.

updating the NXT firmware

Check your version of the NXT firmware by starting BricxCC, connecting to the NXT via USB, and then choosing **Tools ▸ Diagnostics** from the BricxCC menu. In the version label, you should see the Bluetooth and NXT firmware versions. My information says 01.1240/01.028. If the version of the firmware on your NXT is displayed as 01.028, use the RXE programs with the _2 suffix. If the version is 01.05, use the RXE programs with the suffix _1. Close the Diagnostics dialog.

Updating the firmware on your NXT is optional, and **the update will delete all files in the NXT's memory!** So, before you update the firmware, back up your files by copying them from the NXT to your computer using NXT Explorer. To open NXT Explorer, connect the NXT to your computer and select **Tools ▸ NXT Explorer**. As shown in Figure 1-4, the Explorer window is split into three panes. The left pane shows the files in the NXT memory, and the panes on the right (top and bottom) allow you to browse the files on your computer. To make a backup, drag the files from the left pane into an appropriate folder in the right pane.

> Warning! Updating the firmware on your NXT will erase all of its files. Be sure to back up first! In some cases, updating the firmware may cause what has become known as the "clicking brick syndrome." If the update appears to fail and your NXT won't reboot, follow this procedure to fix it: *http://thenxtstep.blogspot.com/2006/06/clicking-brick-syndrome.html.*

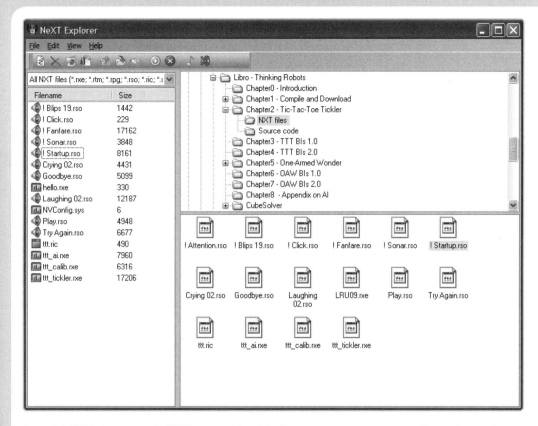

Figure 1-4: NXT Explorer shows the NXT files on the left and the files on your computer on the right. The .rxe files should have the suffix matching your firmware version.

To update your firmware, choose **Tools ▶ Download Firmware**. Use the dialog to browse your computer and locate the firmware file with a *.rxe* extension (in the BricxCC installation directory, or wherever you extracted it) and select it. Click **Open**, and BricxCC should begin to update the firmware. While the firmware is being downloaded, you should see a progress bar, and the NXT should make a soft clicking noise.

Once the update is complete, turn on the NXT then connect it to BricxCC and select **Tools ▶ NXT Explorer**. The NXT's memory should be empty.

NOTE If the update fails and the NXT remains in the clicking mode, try repeating the update procedure. If that doesn't solve your problem, see the warning box on page 6 for information on resetting your NXT.

copying files to the NXT

Once you have the updated version of the firmware on your NXT, copy the files for this book to your NXT. In NXT Explorer, locate the files provided for this book (as listed in Table 1-1) and download them to the NXT by dragging them to the left side panel. You should now be ready to run the programs by selecting them on the NXT's menu.

quick start guide for Mac users

To install the NXT software tools on a Mac, first download the NeXT Tools program for Mac from *http://bricxcc .sourceforge.net/utilities.html*. Next, download and install the latest NXC compiler for Macintosh from *http://bricxcc .sourceforge.net/nbc/* as well as the archive containing the files for this book from *http://tr.benedettelli.com/* or from *http://www.nostarch.com/*, then unzip the downloaded archive.

setting up NeXT Tools

In order to use NeXT Tools, you should have the MINDSTORMS NXT drivers for Mac OS X 10.5 installed, and the NXT turned on and connected to your computer via USB. NeXT Tools is a very useful program, and I encourage you to explore it. Follow these steps to get started:

1. Extract the NeXT Tools program by double-clicking the downloaded archive. A folder named *nxt* should be created.

2. Open NeXT Tools by double-clicking the application icon.

3. Follow the onscreen instructions.

4. Select your NXT name in the Device area, as shown in Figure 1-5, and then click **Select**. (If your NXT name is not displayed, check the connection and be sure that the NXT is turned on.)

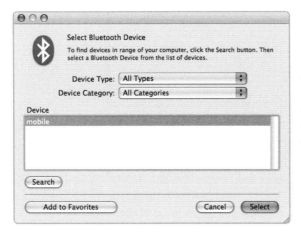

Figure 1-5: Choose your device.

5. In the Select Port dialog, select **USB** from the drop-down list (as shown in Figure 1-6) and click **OK**.

Figure 1-6: Choose USB for the port.

The NeXT Tools main window should now appear, as shown in Figure 1-7.

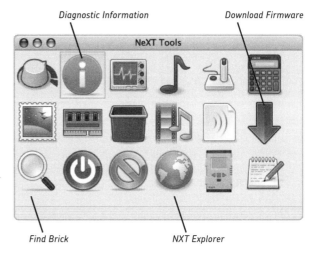

Figure 1-7: The NeXT Tools main window

updating the firmware

Now you need to make sure that the NXT has the correct firmware version.

NOTE Updating your firmware will delete all files from the NXT. Be sure to back up any files on your NXT before performing a firmware update!

1. In the NeXT Tools window, click the Diagnostic Information icon to check the version of the NXT firmware. If the version of the firmware is 01.028, you must use the .rxe programs with the _2 suffix. If the version is 01.05, you must use the RXE programs with the suffix _1. Close this window.

2. If you want to update your firmware to version 1.28, click the download icon, browse to the file *lms_arm_nbcnxc_128.rfw*, and click **Open** to start the firmware update process. While the firmware is being updated, you should see a progress bar, as shown in Figure 1-8.

Figure 1-8: The Firmware Download progress bar

3. Once the firmware update completes, the NeXT Tools main window should reappear, but some icons will be grayed out because the NXT has lost the connection. Click the Find Brick icon, choose the USB port again, and click **OK**.

copying files to the NXT

Once you have the updated version of the firmware, you can place the files for this book on your NXT. In the NeXT Tools window, click the NeXT Explorer icon. An empty window showing your NXT memory should appear (like the one shown in Figure 1-9).

To copy all of the necessary files to the NXT, click the Download Firmware icon (the downward-pointing arrow). Browse to the folder containing the downloaded files for your robot (as listed in Table 1-1), select all of the files, and then click **Open**. The memory panel should fill with the selected files, as shown in Figure 1-9. Now you are ready to run all the programs by selecting them from the NXT's menu.

compiling NXC programs using the NXC compiler for Mac OS X

The NXC compiler for Mac OS X exists only as a command-line version. In order to compile NXC programs on Mac OS X, you will need the appropriate compiler, as noted at the beginning of this quick start guide.

Here's an example of how to create a simple NXC source file, compile, and download it to the NXT.

Open TextEdit on your Mac, and then create a text file with the following contents:

```
task main ()
{
    TextOut (10,LCD_LINE4,"Hello, World!");
    Wait (SEC_2);
}
```

This trivial NXC program prints "Hello, World!" on the screen, waits for two seconds, and then ends.

Save the file with the name *hello.nxc* in the *nxc* folder (the one that contains the compiler), then open the Terminal

Filename	Size
Goodbye.rso	5099
Laughing 02.rso	12187
! Blips 19.rso	1442
! Click.rso	229
! Fanfare.rso	17162
Try Again.rso	6677
Crying 02.rso	4431
! Sonar.rso	3848
ttt.ric	490
ttt_ai.rxe	7960
Play.rso	4948
ttt_calib.rxe	6316
ttt_tickler.rxe	17206
LRU09.rxe	15202
NVConfig.sys	6

Figure 1-9: The NeXT Explorer window should look like this after downloading all the files for this book. The rxe program files should have the suffix matching your firmware version.

program (you can find it in the Applications folder of your Mac), and navigate to the *nxc* folder using the **cd** command:

```
cd /your_path/nxc
```

To get help with the NXC compiler, enter the following:

```
./nbc -help
```

To compile the *hello.nxc* file into the *hello.rxe* binary file, enter the following:

```
./nbc hello.nxc -O=hello.rxe
```

To download the compiled binary file *hello.rxe* to the NXT, turn on the brick, connect it to your Mac, and enter:

```
./nxtcom -U hello.rxe
```

The NXT should beep to confirm the successful file transfer. Once the transfer is complete, you should be able to run the *hello.rxe* program by selecting it from the NXT's menu.

summary

In this chapter, Windows users learned how to compile NXC source code and how to use the BricxCC to open and download the programs provided with this book to the NXT. Macintosh users learned how to do the same, using the NeXT Tools and NXC command-line compiler.

2

the TTT Tickler:
a tic-tac-toe player

Raise your hand if you have ever spent time playing tic-tac-toe during some boring and never-ending school lesson! Just draw four crossed lines on any surface (whether it's wastepaper, an exercise book, or even a desk), and the game board is ready!

In this chapter, you'll learn the perfect strategy for tic-tac-toe, as well as how to play the game with the Tic-Tac-Toe Tickler (TTT Tickler), a LEGO MINDSTORMS NXT robot that will challenge you at every turn.

You'll find building instructions for two versions of this robot in the following chapters. Chapter 3 contains instructions for use with the original LEGO MINDSTORMS NXT set (8527), and Chapter 4 contains building instructions for the LEGO MINDSTORMS NXT 2.0 set (8547). You can see the two versions of the robot in Figures 2-1 and 2-2.

The two versions of the robot work exactly the same way. The differences are aesthetic, due to the varying assortment of parts contained in the two sets. For example, in the NXT 2.0 version, you can use the Zamor magazine to hold the balls; in the NXT 1.0 version, you need to build a magazine using the spare parts available.

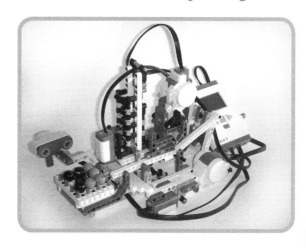

Figure 2-1: The TTT Tickler built with the NXT 8527 set. (Notice the glass marbles used as marks—dark for you and light for the robot.)

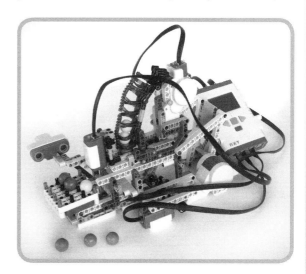

Figure 2-2: The TTT Tickler built with the NXT 8547 set

make the perfect game

The game of tic-tac-toe is also known as noughts and crosses or hugs and kisses (in Italian, *filetto* or *tris*). The two players, O and X, alternatively draw their mark on a 3-by-3 grid. The player who places three marks in a horizontal, vertical, or diagonal row wins the game. Figure 2-3 shows an example in which player O starts and wins the game.

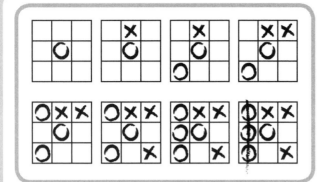

Figure 2-3: An example of a tic-tac-toe match

Once you've played tic-tac-toe a few times, you'll soon discover that when both you and your opponent play with the optimal strategy, the game ends in a draw.

NOTE The simplicity of tic-tac-toe makes it ideal as an educational tool to explain the basic concepts of *combinatorial game theory* and *artificial intelligence*. These concepts are the subject of study of a branch of information technology, and they can be used to instruct a computer to play a game. For more on this topic, see the appendix.

Have you ever wondered how to play a perfect tic-tac-toe match? The following are the rules to win (or at least to end in a draw). These are the same rules that I've programmed into the TTT Tickler robot to make it a flawless player! These rules assume that you are player O. Every time it's your turn, you should choose one of the following moves, depending on the state of the game. The rules are listed in order of decreasing priority.

1. **Try to win** If you have two marks in a row, put the third mark in an empty space to get three in a row and win the game (as shown in Figure 2-4).

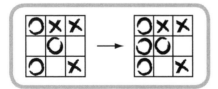

Figure 2-4: Try to win.

2. **Block the opponent** If your opponent has two X marks in a row, place your mark to block him (as shown in Figure 2-5).

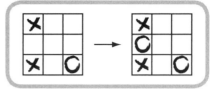

Figure 2-5: Block your opponent.

3. **Fork** Create an opportunity to win in two ways by making a fork (as shown in Figure 2-6).

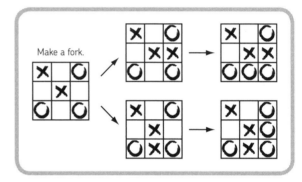

Figure 2-6: Make a fork.

4. **Block your opponent's fork** If there's a configuration where your opponent can fork, block that attempt. For example, you might try to put two marks on a row to force your opponent to block you (as shown in Figure 2-7).

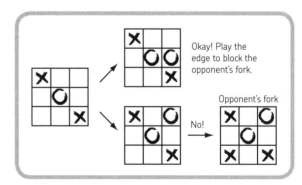

Figure 2-7: Block your opponent's possible fork.

When none of the above situations apply, here is the basic game play (as shown in Figure 2-8):

* Place your mark in a square in the center of the board.
* If your opponent places a mark in the corner, mark the opposite corner in the same diagonal.
* Place a mark in an empty corner square.
* Place a mark in an empty edge square.

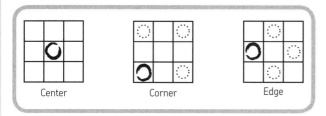

Center   Corner   Edge

Figure 2-8: Play in the center, in the corner, or in an edge square.

When you play against the TTT Tickler, you are player O, and you make the first move; the robot is player X.

While it may seem like you can choose among nine possible places on the board at the start of the game, rotations and symmetries lead to only three possibilities: center, corner, or edge. In fact, when the board is empty, every corner or edge has the same strategic importance. As shown in Figure 2-9, placing your first mark in the center square offers you four possible winning configurations; more than any other option. Placing your first mark in a corner offers three potential winning configurations, and placing your first mark on an edge offers only two possible winning configurations.

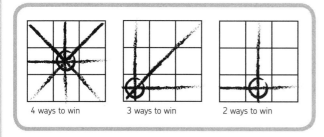

4 ways to win   3 ways to win   2 ways to win

Figure 2-9: Three possible choices for your initial move

If you place your first mark in a corner, the robot (player X) will put its mark in the center square. If you place your first mark in the center square, the robot will place its mark in a corner square. If, however, you open with an edge mark, the robot responds by marking the center square (though marking an adjacent corner or an opposite edge would also be a good move). After the opening moves, the robot follows the preceding priority list to force a draw, or a win if you make a mistake!

introducing the TTT Tickler

There are many LEGO MINDSTORMS robots that play tic-tac-toe, but none is like the one you are going to build.

NOTE Mario Ferrari created his TTT robot in 1999 using the RCX brick. You can see his robot by going to *http://www.marioferrari.org/ttt.html*. More recently, in 2006, Bryan Bonahoom built the NXT robot W.O.P.R. (*http://mindstorms.lego.com/MeetMDP/BBonahoom.aspx*).

The TTT Tickler is different because it's very compact. It uses one motor to rotate the board, another to move the arm linearly on the board, and a third to drop the balls on the board.

The robot's rotating board has nine spaces where you and the robot can place balls of different colors. When playing with the original NXT version, you can use dark and light glass marbles (five of each); with the NXT 2.0 version, you can use the Zamor Spheres included in the set. You should use green and blue balls (which appear dark to the sensor), and fill the robot's magazine with the yellow and red ones (which appear light to the sensor). As you should be able to see in Figure 2-10, the robot can reach all cells on the board by rotating the board in 45-degree increments and sliding the arm in and out.

The LEGO color or light sensor on the end of the robot's arm should detect where you've placed your ball. The robot drops its balls on squares through a chute by pushing them out of the magazine with a cam mechanism driven by the vertical motor.

user manual

In this section, you will learn how to play with the TTT Tickler. I've provided three programs for use with your robot: one to play tic-tac-toe using the NXT's screen, one to play with the robot directly, and one to recalibrate the robot's arm.

Because these are quite complicated programs written in the NXC language, I won't take you through writing them in this book; however, you can download them all from *http://tr.benedettelli.com/* or *http://www.nostarch.com/*. (Chapter 1 contains instructions on how to download files and programs to the NXT.) The following files are used:

> **Compiled programs** *ttt_ai_1.rxe*, *ttt_tickler_1.rxe*, and *ttt_calib_1.rxe* (firmware 1.05); *ttt_ai_2.rxe*, *ttt_tickler_2.rxe*, and *ttt_calib_2.rxe* (firmware 1.28)

> **Sounds** *! Blips 19.rso*, *! Click.rso*, *! Sonar.rso*, *Crying 02.rso*, *! Fanfare.rso*, *Goodbye.rso*, *Laughing 02.rso*, *Play.rso*, *Try Again.rso*, and *Yes.rso*

> **Image** *ttt.ric*

playing against the robot

To play against the robot, select the *ttt_tickler* program and press the orange button on the NXT. The robot should reset the board position, its arm, and the ball-ejection mechanisms. It will ask if you want to calibrate the sensor with the room based on the light around you (*ambient* light). Select **Yes** the first time you use the robot and then again every time the light conditions change. I suggest calibrating the light sensor often, ideally every time you turn on the NXT Brick.

Continue the calibration by following the instructions on the screen, which tell you to put a dark ball (a blue or green Zamor Sphere or a dark marble) on the central square and press the NXT's orange button. When you do, the robot measures the ball color and stores it in a configuration file called *ttt_s.cfg*. Continue to follow the instructions on the screen. Finally, remove all balls from the board and make sure that the robot has at least four balls in its magazine. Press the NXT's orange button again when you are ready to play.

You can choose to play against a Novice, a Normal, or an Expert opponent. Beating the Novice is quite easy, since it plays randomly. You can beat the Normal opponent using the strategy described in "Make the Perfect Game" on page 11, but it is up to you to discover its Achilles' heel! There is no way to beat the Expert opponent, but if you play well, you can tie.

Once you have chosen the robot's skill level, the robot waits for you to make the first move. It automatically detects your move by checking for the presence of your hand with the ultrasonic sensor, so make sure that when you put the ball on the board the robot can see your hand. (You'll know that it has seen your hand because the NXT will beep.)

Each time the TTT Tickler takes a turn it scans the board for open squares, skipping already occupied squares to save time.

Once the robot has scanned the board, it computes the best move and drops a ball from its magazine onto the board. This cycle repeats until the game ends, at which point you can choose to play another game or simply end. (If you have thrashed the robot and choose to stop playing, it is still a good sport and will see you off with a smile!)

calibrating the robot

When the robot fails to place the sensor correctly or drops the ball in the wrong place, you should run the *ttt_calib* calibration program. You should calibrate the robot arm **only** if it often fails to drop balls into their correct places.

NOTE Calibration is a delicate procedure that normally should be avoided. If you perform calibration incorrectly, the robot's arm may not move correctly. To restore the default calibration settings, simply delete the *ttt_a.cfg* file on the NXT.

To delete an existing *ttt_a.cfg* configuration file, start the *ttt_calib* program, release the NXT's orange button, and press and hold it again immediately. The NXT should confirm the file deletion both on the screen and by making a sound. Alternatively, you can delete this file using the NXT Explorer (BricxCC in Windows) or the NeXT Explorer tool (NeXT Tools on a Macintosh).

Now, to begin the calibration, follow the onscreen instructions. **Rotate the knob on the shaft of the arm's motor by hand** (attached to OUT B) to slide the arm, saving each arm position by pressing the orange button as shown in Figure 2-10. It's very important to not move the arm directly or the motor angle may be recorded incorrectly; turn the knob instead.

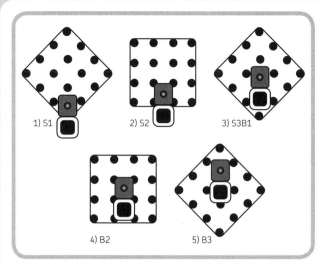

Figure 2-10: The robot arm must be positioned by turning the knob on the motor by hand in order to perform the calibration.

The arm positions are as follows:

S1 (sensor position 1) Move the arm so that the sensor spot is in the corner.

S2 (sensor position 2) Move the arm so that the sensor spot is on the edge.

S3B1 (sensor position 3, ball position 1) Move the arm so that the sensor spot is in the center and the ball can be dropped in the corner.

B2 (ball position 2) Move the arm so that the ball can be dropped on the edge.

B3 (ball position 3) Move the arm so that the ball can be dropped in the center.

At the end of the calibration procedure, the robot's moving parts should reset, and the NXT should display a summary of the new positions, as stored in the *ttt_a.cfg* file. At startup, the *ttt_tickler* program looks for this configuration file and, if it exists, loads its settings instead of the default ones.

playing the screen-based tic-tac-toe game

You can also play tic-tac-toe on the NXT brick directly, without using the robot. Start the screen-based game by selecting *ttt_ai* and pressing the NXT's orange button. The game board should be displayed on the NXT screen.

When the program starts, you can choose the opponent's skill level. To select a square on the board, move the cursor using the NXT arrow buttons and confirm your move by clicking the orange button. The NXT should compute its move, update the screen, and repeat the cycle until the game ends.

summary

In this chapter, you learned the optimal strategy for playing tic-tac-toe, as programmed into the TTT Tickler, and how to use and configure your robot. Next, you will build the robot.

3

building the TTT Tickler with the original NXT set

In this chapter you will build the TTT Tickler with the parts from the 8527 NXT set. Use some glass marbles as marks for the game: you will play with five dark-colored marbles and you should fill the robot's magazine with five light-colored ones.

bill of materials

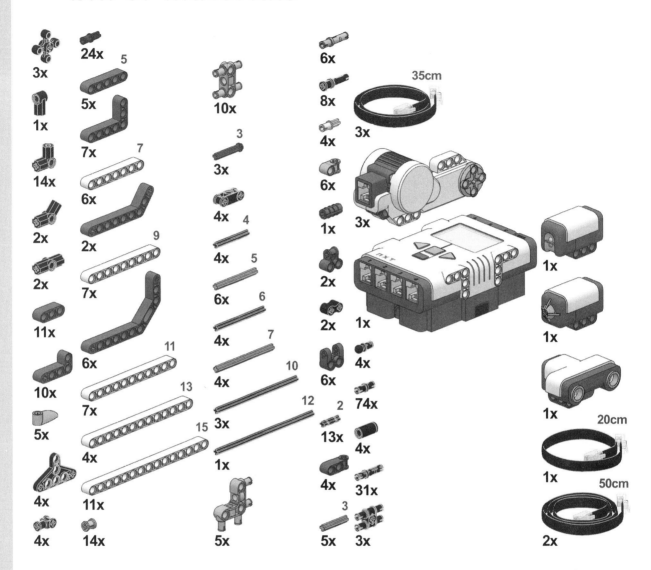

main assembly

1

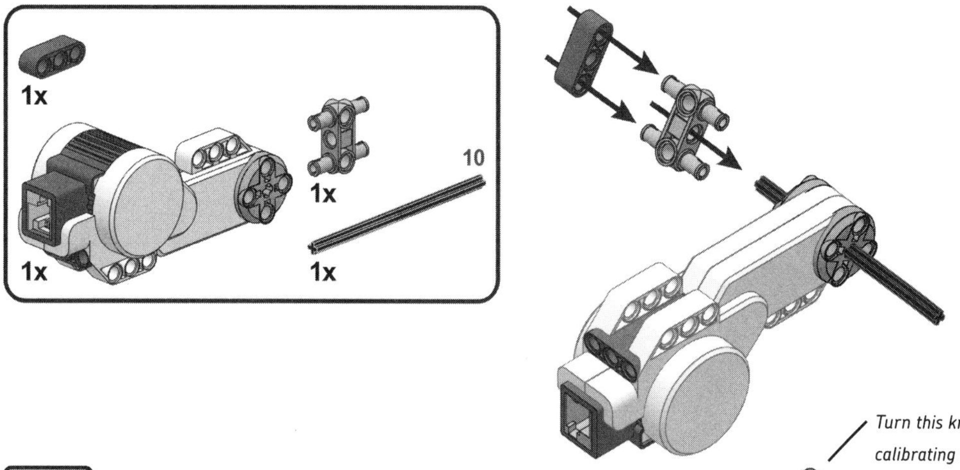

1x

1x 1x 10 1x

Turn this knob by hand when calibrating the robot.

2

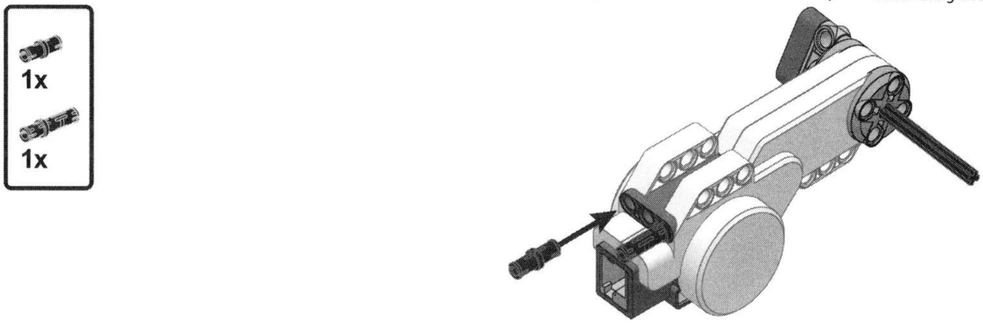

1x

1x

3

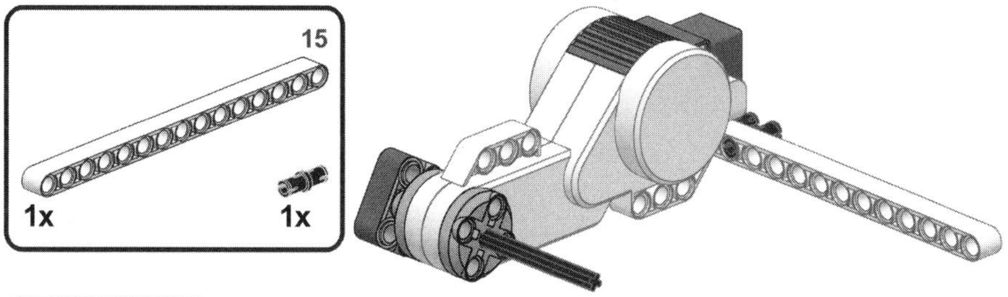

15

1x 1x

4

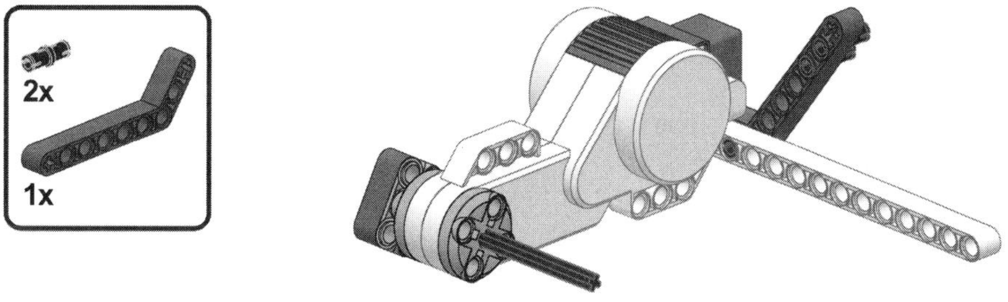

2x

1x

5

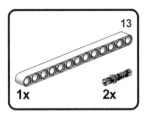

1x   2x

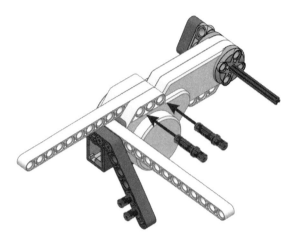

6

1x   1x

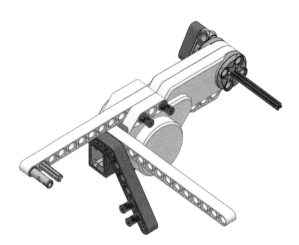

7

2x

1x

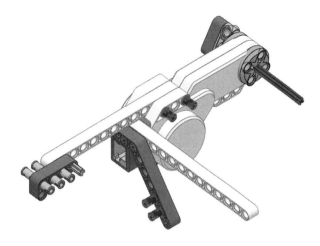

8

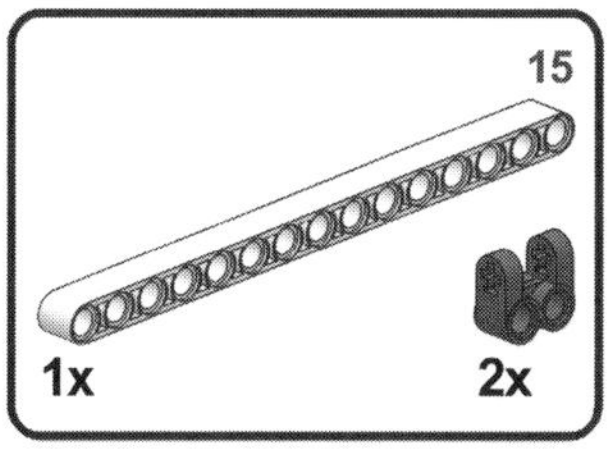

15

1x   2x

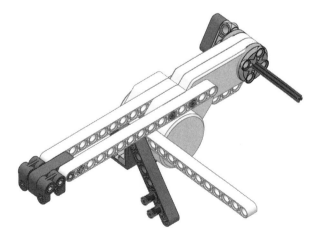

9

2x

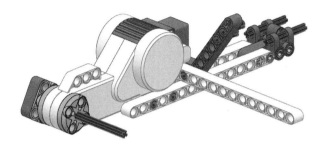

10

1x

1x

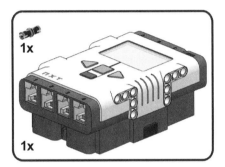

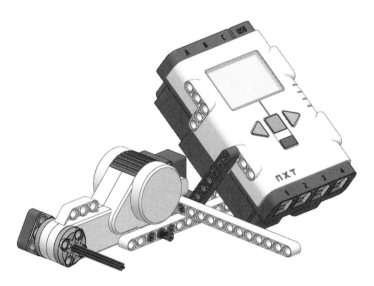

11

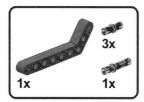

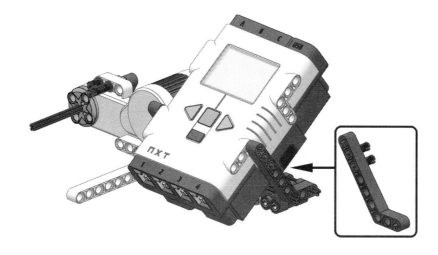

12

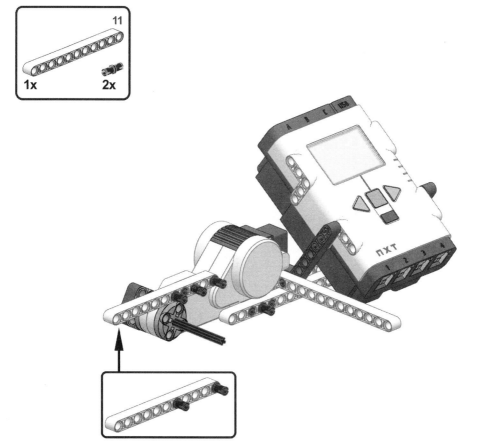

13

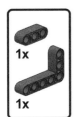

1x
1x

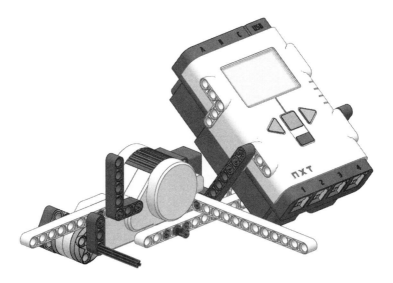

14

1x
1x
1x
2x

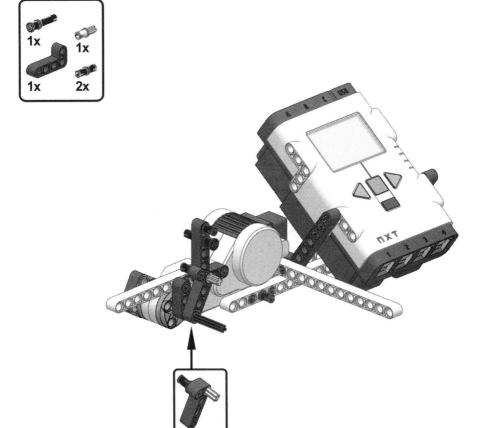

15

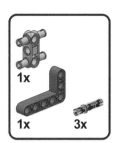

1x
1x   3x

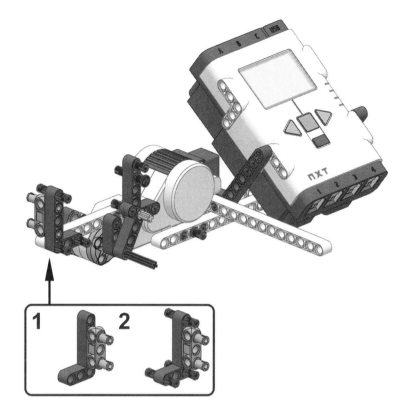

1 2

base

1

1x
1x   3x   15

2

1x

3

1x
1x

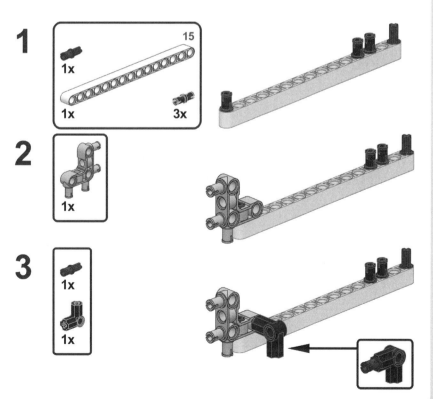

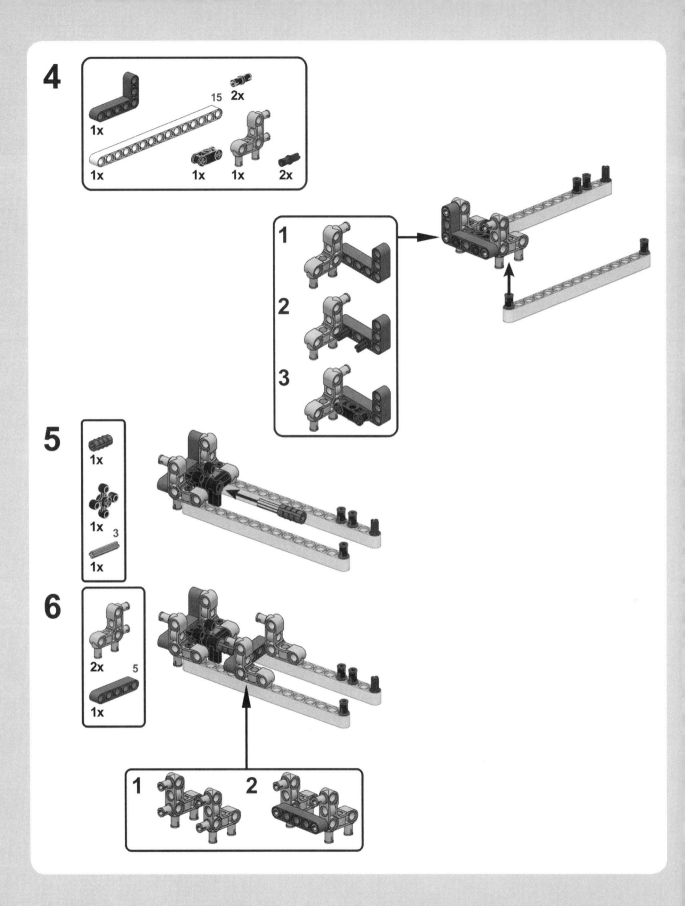

7

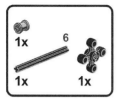

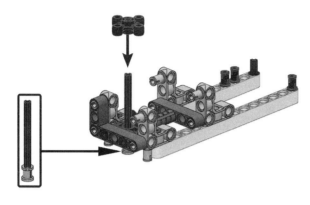

8

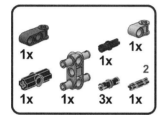

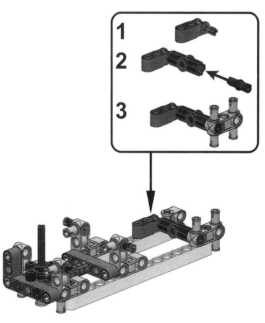

9

12 1x 1x

10

1x **5** 1x

11

1x **15** 2x 1x

1x 1x 1x 2x

1 **2** **3**

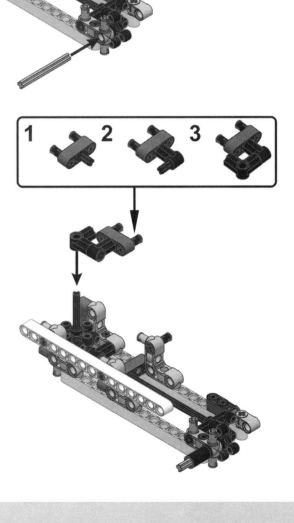

12

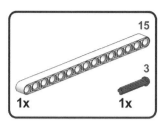

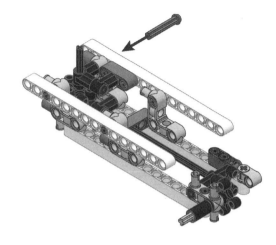

13

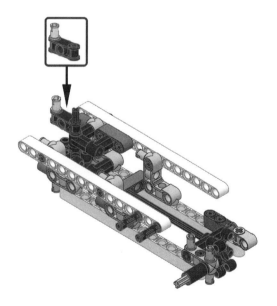

14

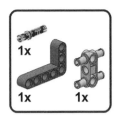

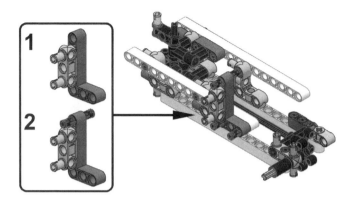

Return to the main assembly.

16

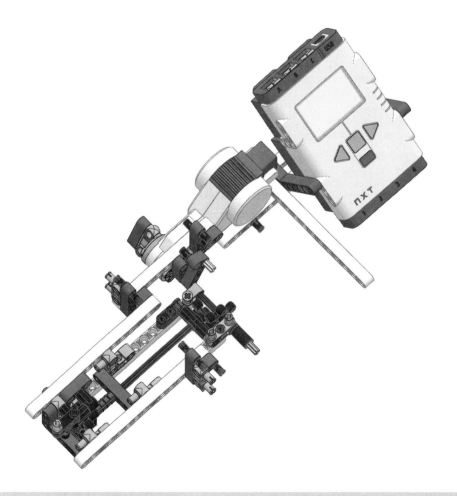

lever

1

13

1x 1x

2

1x

1x

1x

3

1x

4

1x

13

1x

5

1x

2x

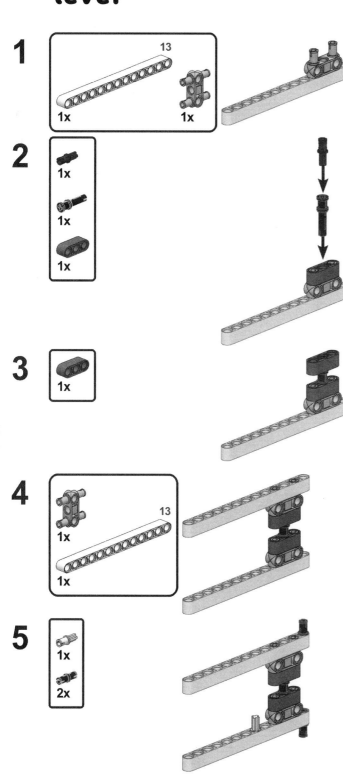

Return to the main assembly.

17

18

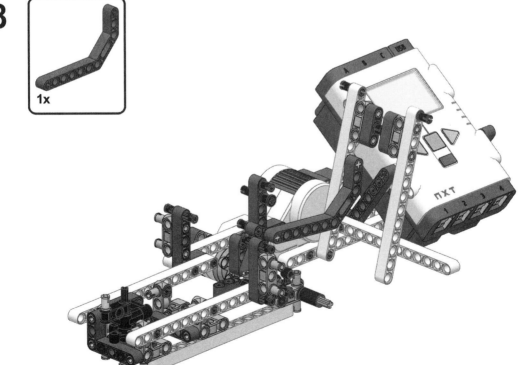

board motor

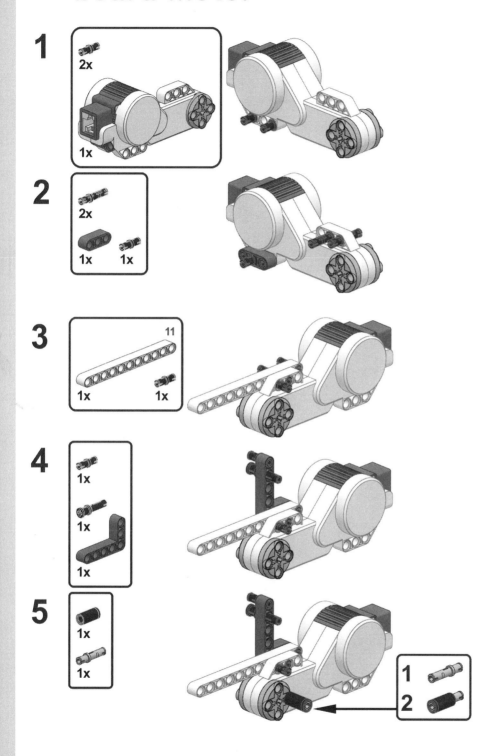

1 2x 1x

2 2x 1x 1x

3 11 1x 1x

4 1x 1x 1x

5 1x 1x

1
2

Return to the main assembly.

19
1x

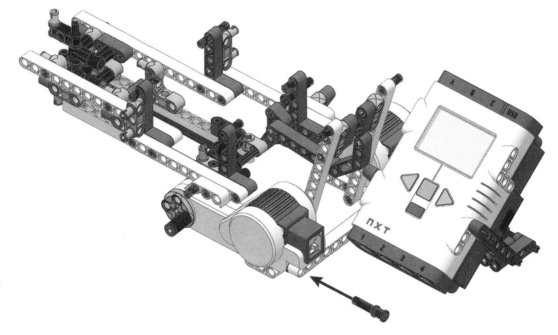

20

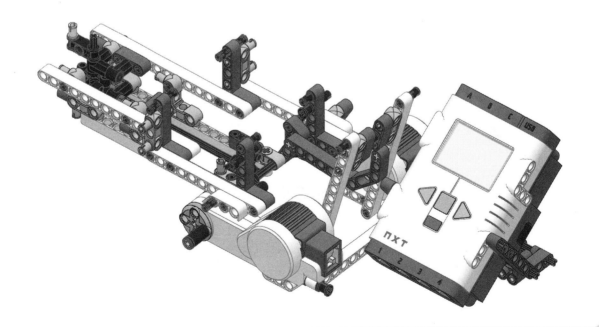

touch sensor

1

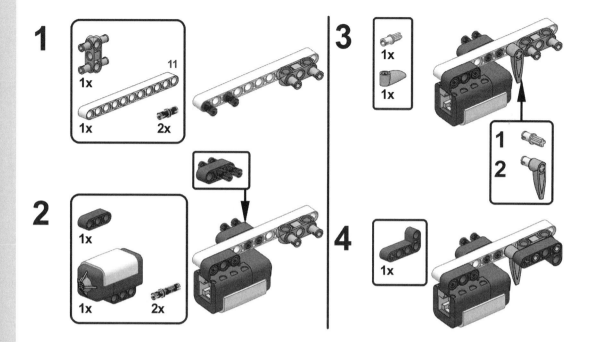

2

3

4

Return to the main assembly.

21

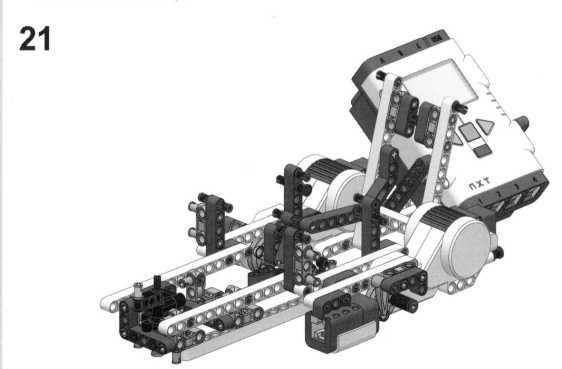

22

15
2x

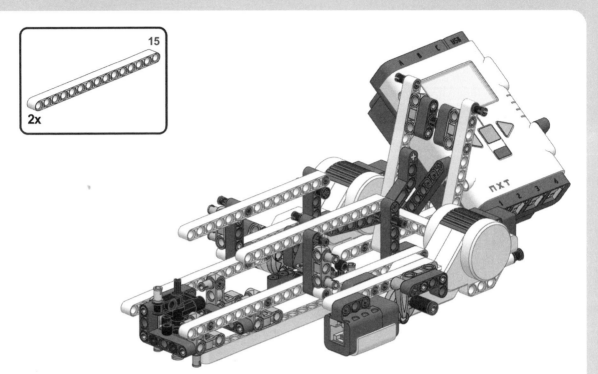

arm

1

1x
1x   2x

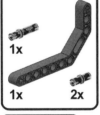

2

1x
7
1x

3

7
1x

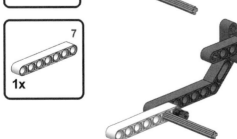

4

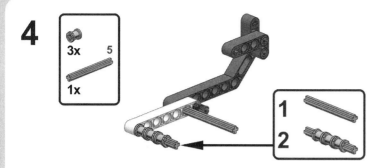

cursor guide

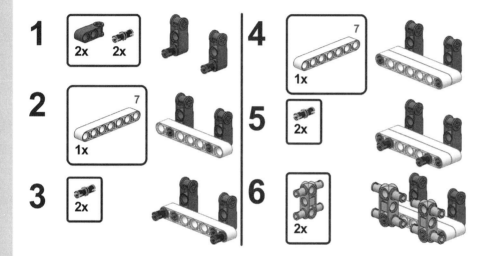

Return to the arm subassembly.

5

6

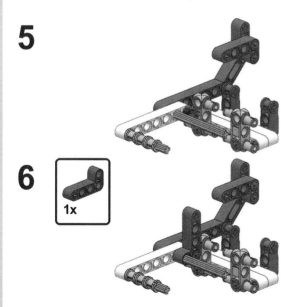

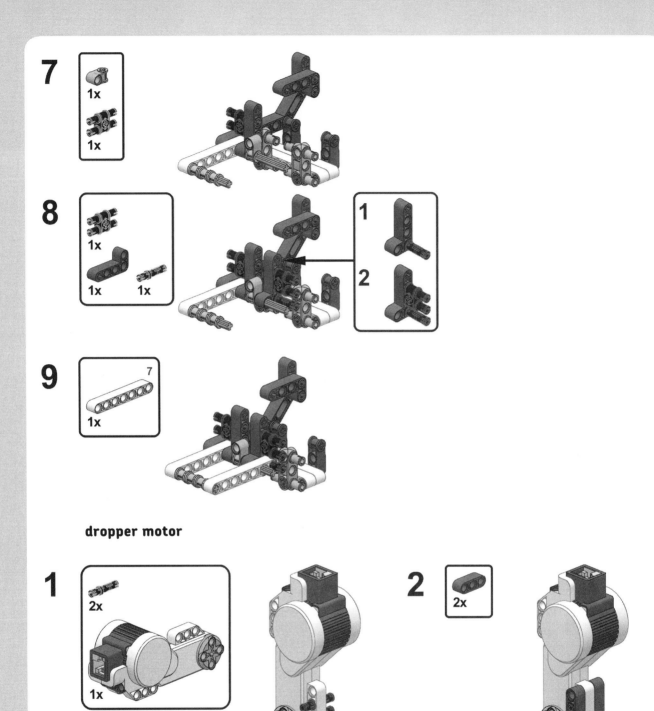

dropper motor

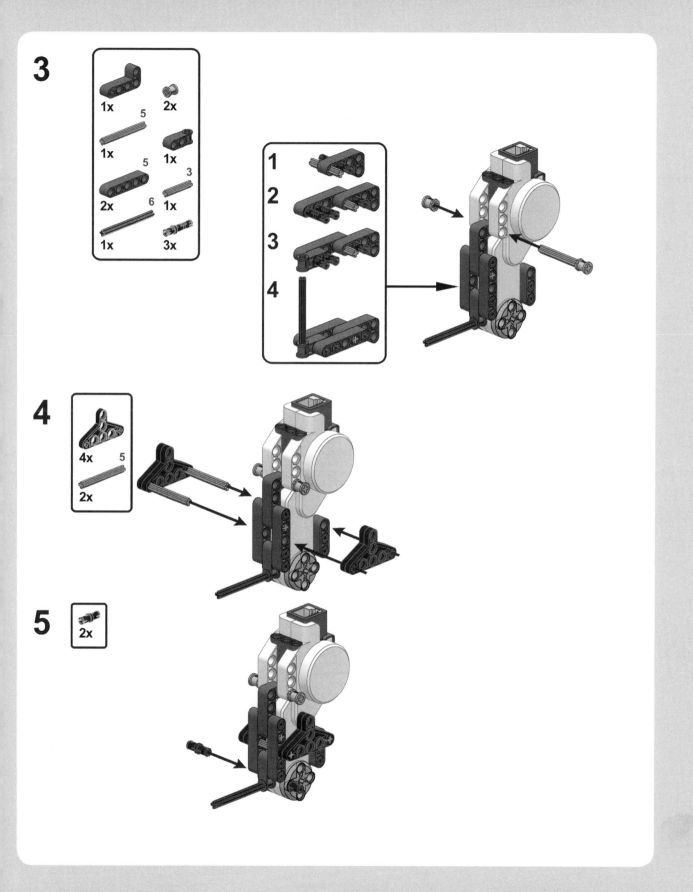

6

5

2x

7

3

2x

1x

8

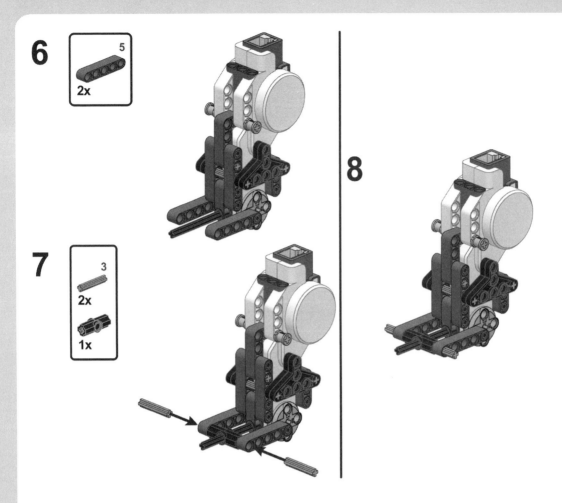

Return to the arm subassembly.

10

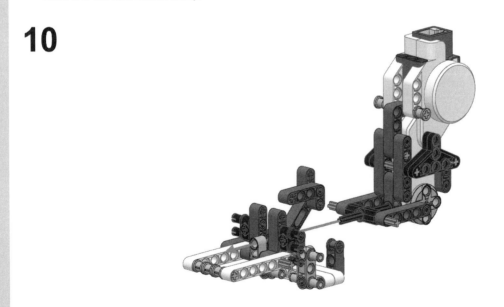

11

1x
1x 2x

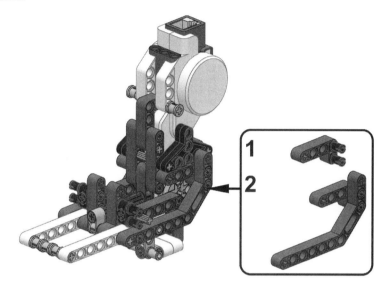

1
2

12

2x

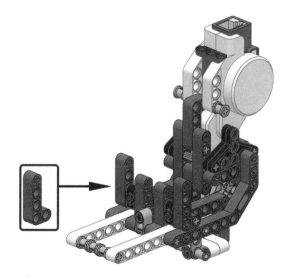

marbles magazine

1

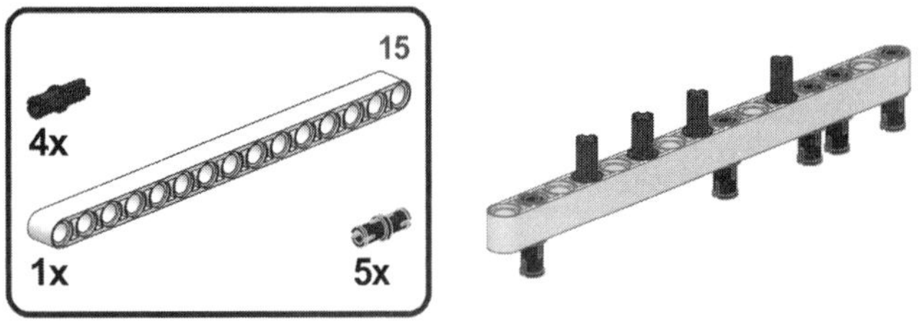

15
4x
1x
5x

2

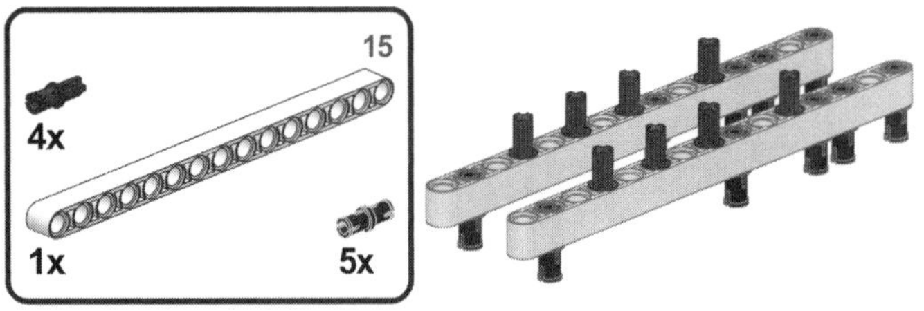

15
4x
1x
5x

magazine cage

1

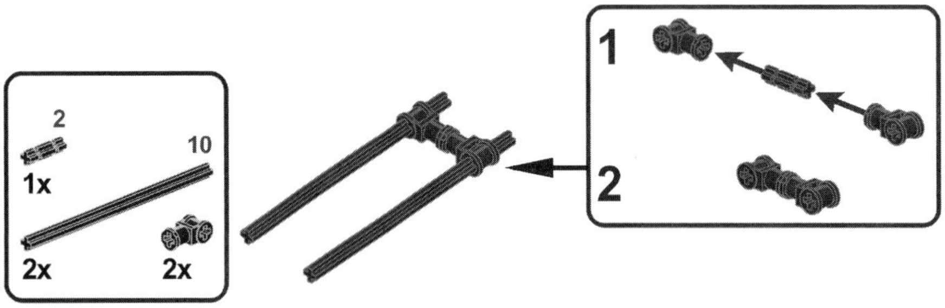

2
1x
10
2x
2x

1
2

2

2x

3

2x

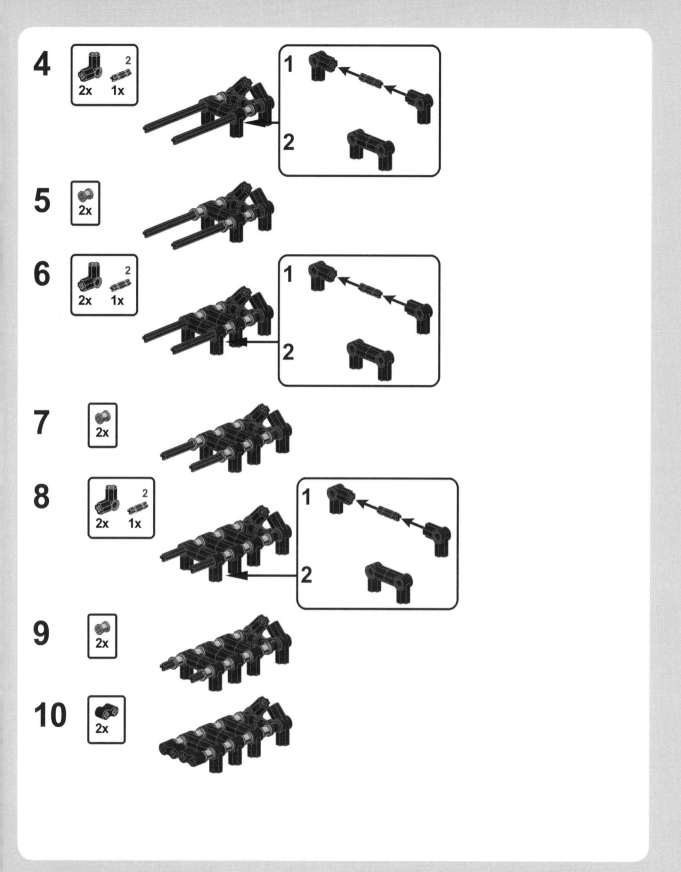

Return to the marbles magazine subassembly.

3

4

5

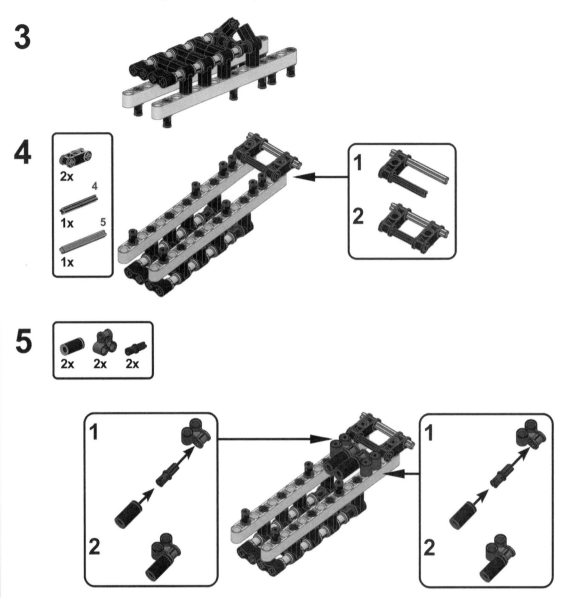

magazine back cage

1

2x   2x   1x

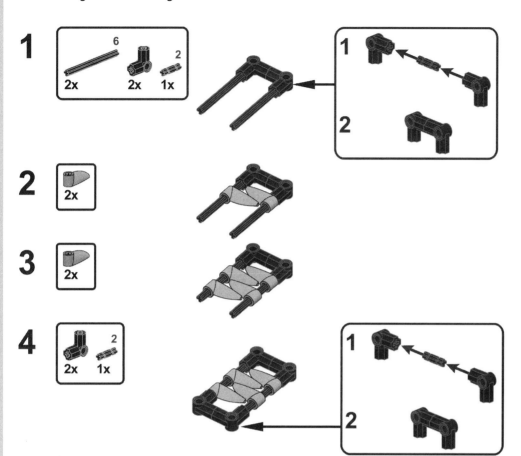

2

2x

3

2x

4

2x   1x

Return to the marbles magazine subassembly.

6

light sensor

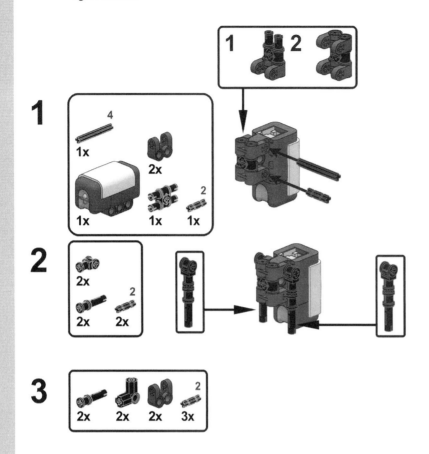

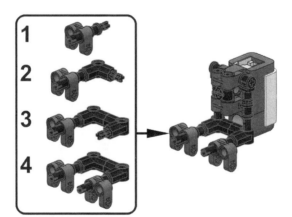

Return to the marbles magazine subassembly.

7

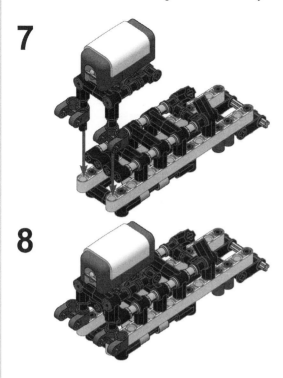

8

Return to the arm subassembly.

13

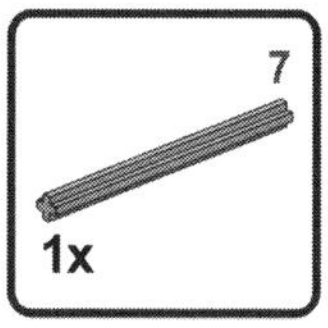

1x

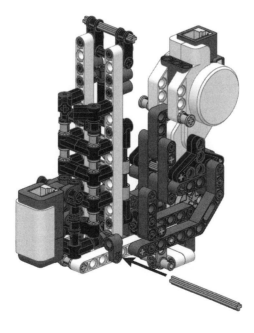

14

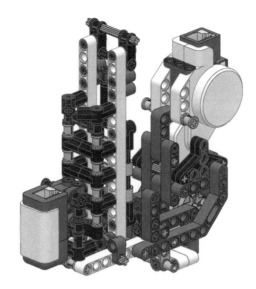

Return to the main assembly.

23

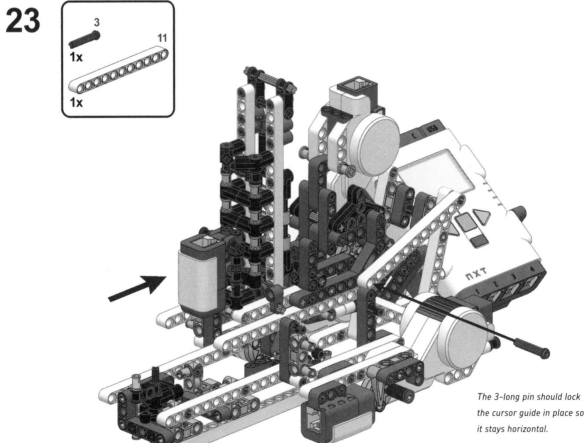

The 3-long pin should lock the cursor guide in place so it stays horizontal.

24

3

11

1x

1x

flip!

The 3-long pin should lock the cursor guide in place so it stays horizontal.

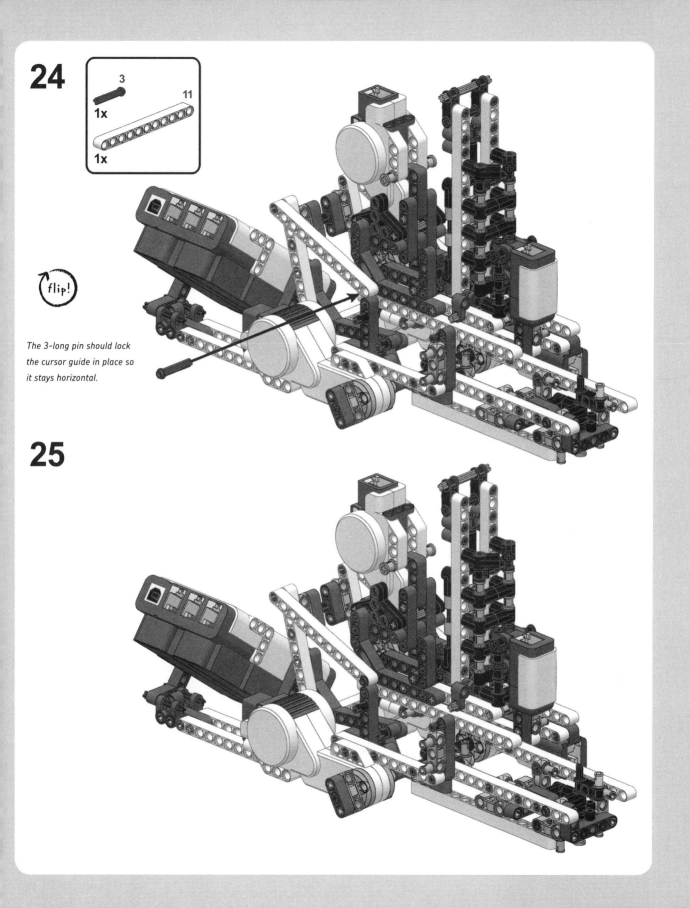

25

plate

1

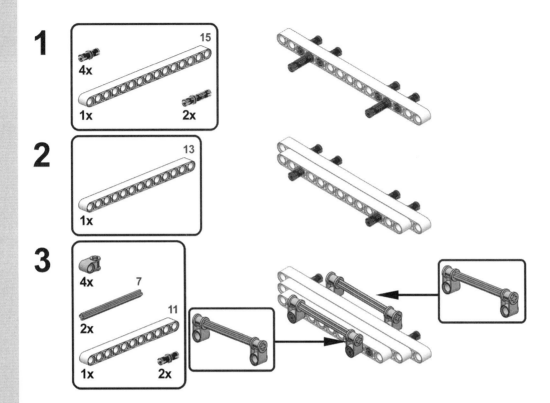

2

3

Return to the main assembly.

26

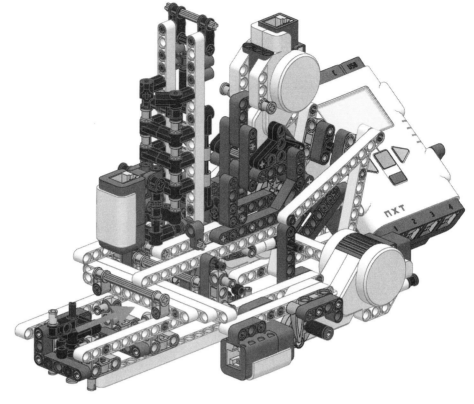

27

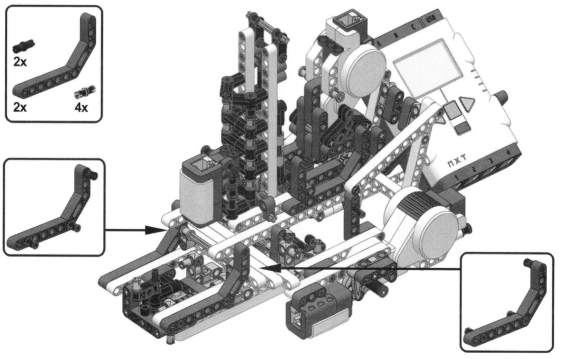

board

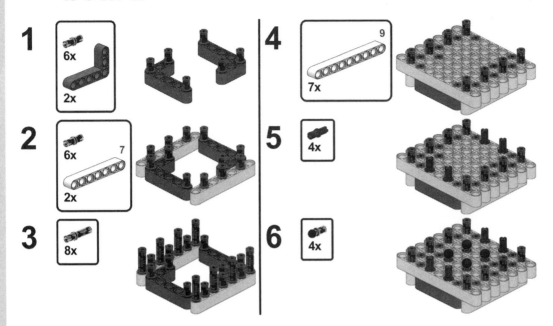

1 6x 2x

2 6x 2x — 7

3 8x

4 7x — 9

5 4x

6 4x

Return to the main assembly.

28

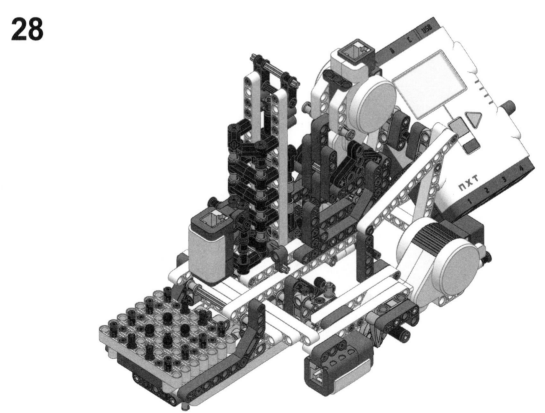

ultrasonic sensor

1

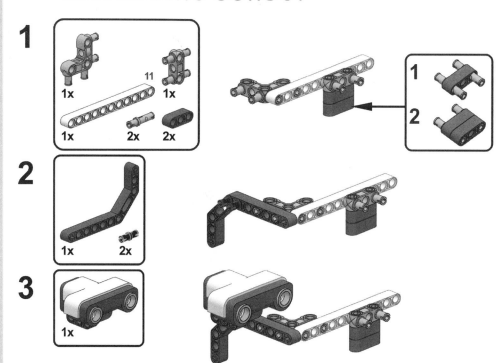

2

3

Return to the main assembly.

29

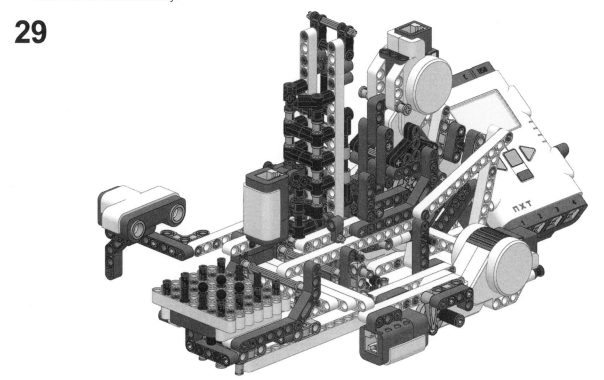

30

20cm

1x

Attach the arm motor to OUT B
with a 20cm (8") cable.

31

35cm

1x

Attach the dropper motor to
OUT A with a 35cm (14") cable.

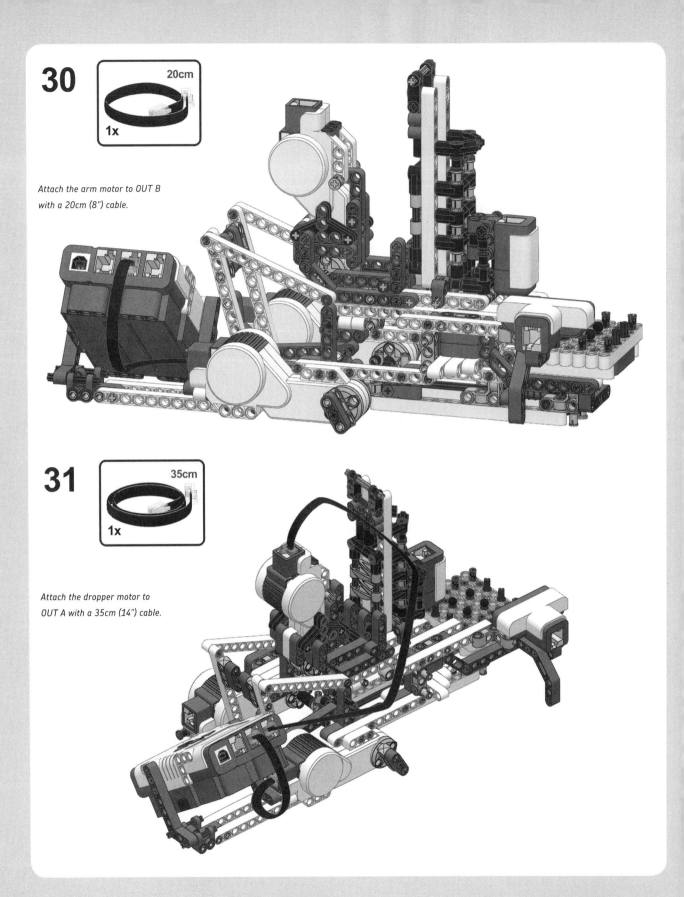

32

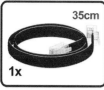

Attach the touch sensor to IN 2
with a 35cm (14") cable.

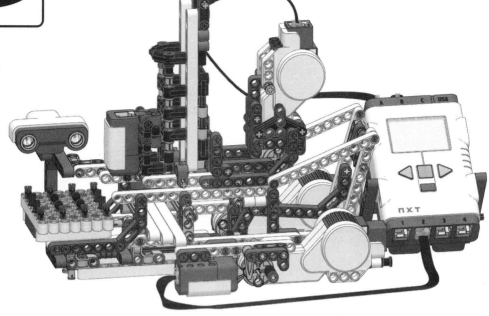

33

Attach the ultrasonic sensor to
IN 3 with a 50cm (20") cable.

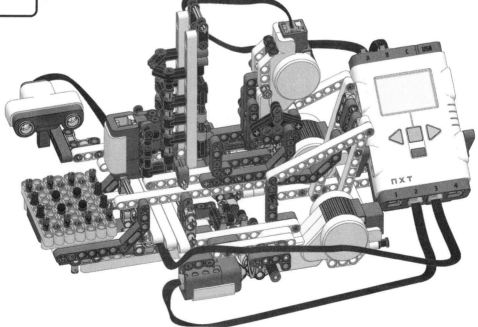

34

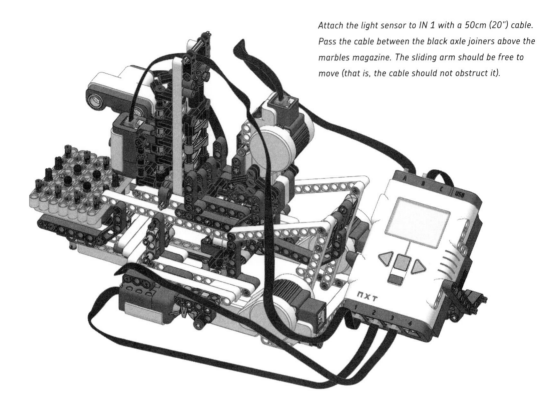

Attach the light sensor to IN 1 with a 50cm (20") cable. Pass the cable between the black axle joiners above the marbles magazine. The sliding arm should be free to move (that is, the cable should not obstruct it).

35

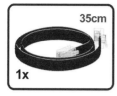

Connect the board motor to
OUT C with a 35cm (14") cable.

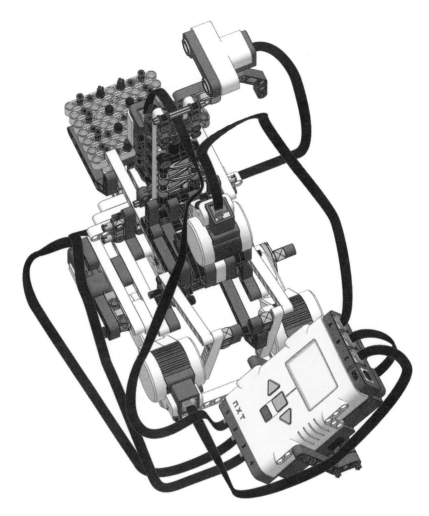

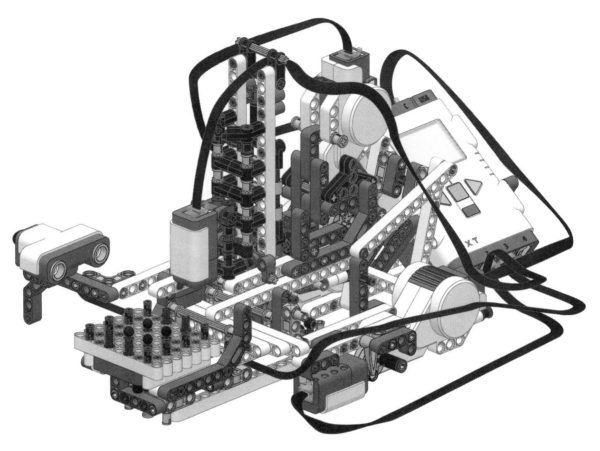

36

The TTT Tickler is ready to play!

building the TTT Tickler with the NXT 2.0 set

In this chapter you will build the TTT Tickler with the parts from the 8547 NXT 2.0 set. Use the Zamor Spheres as marks for the game: you should play with green and blue spheres (which appear dark to the sensor), and fill the robot's magazine with the yellow and red ones (which appear light to the sensor).

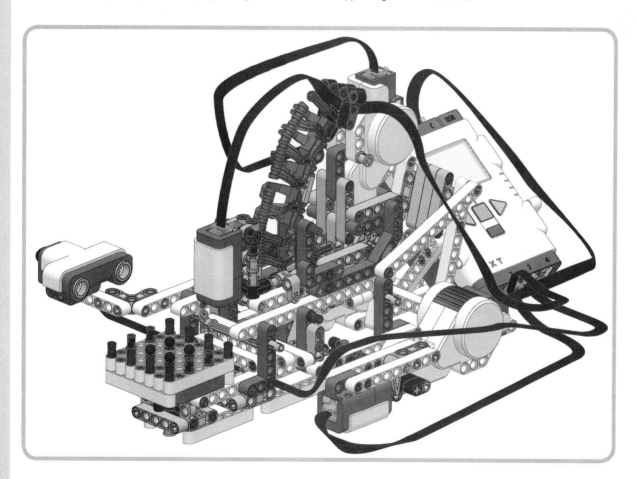

bill of materials

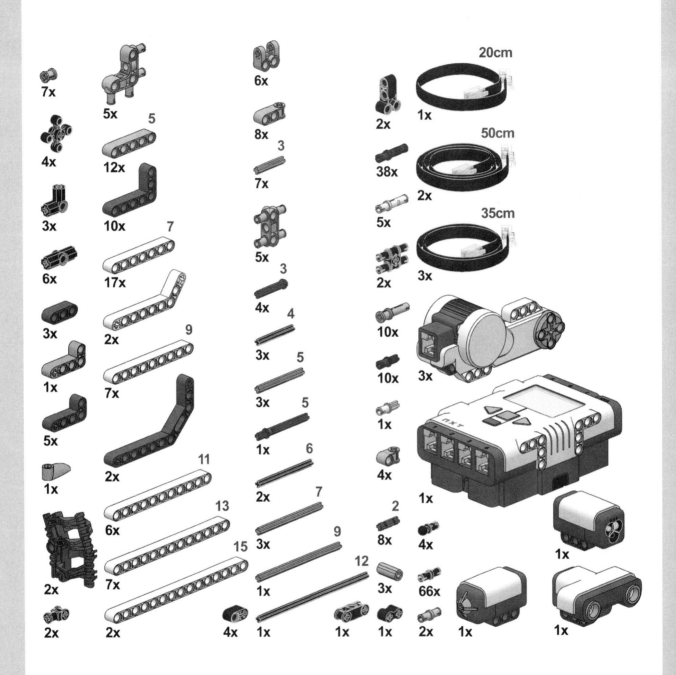

main assembly

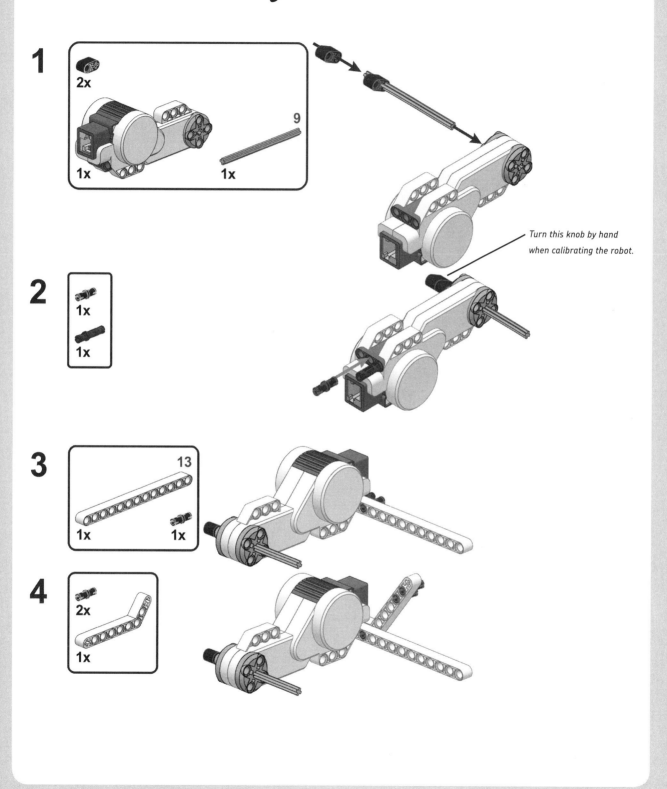

1

2x

1x   9   1x

*Turn this knob by hand
when calibrating the robot.*

2

1x

1x

3

13

1x   1x

4

2x

1x

5

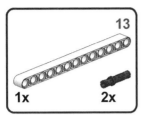

1x   2x

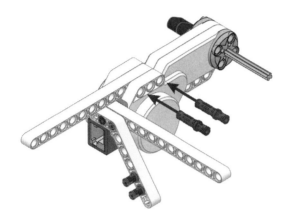

6

2x

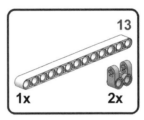

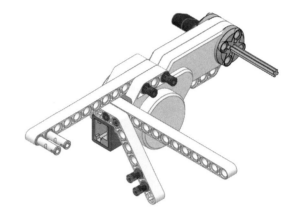

7

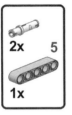

2x   5

1x

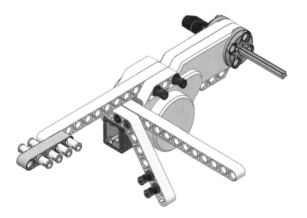

8

13

1x   2x

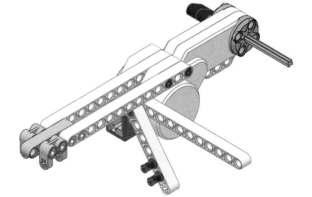

9

4

2x

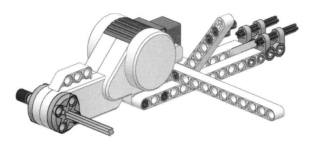

10

1x

1x

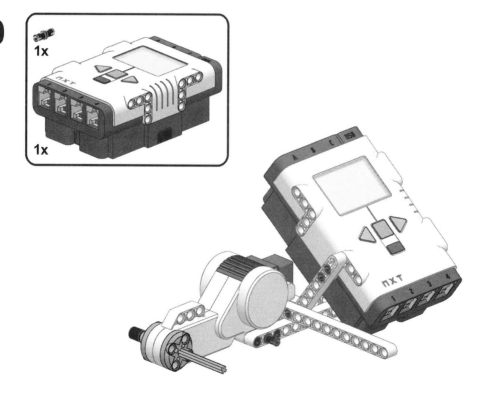

11

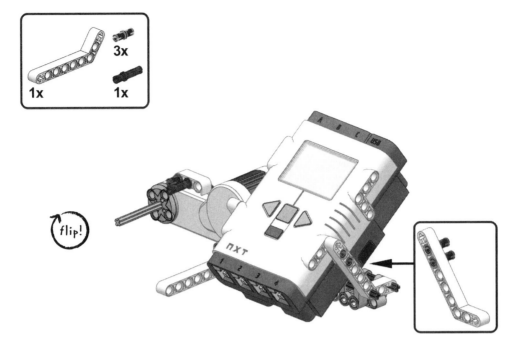

12

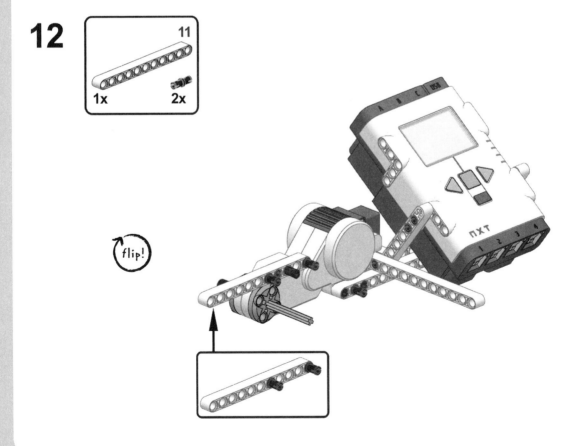

13

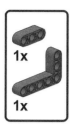

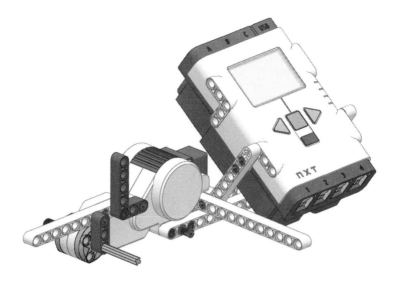

14

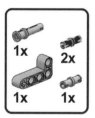

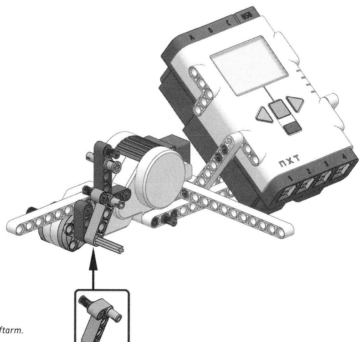

Use an orange 2 × 4 liftarm.

15

2x 1x
1x 3x

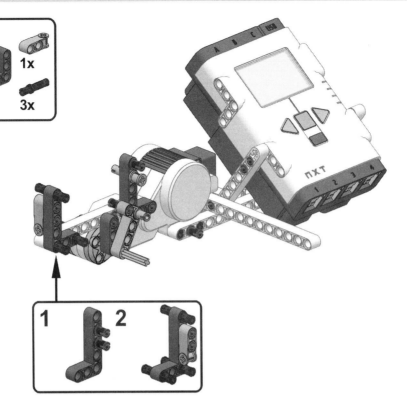

1 2

base

1

7
1x 1x

2

2x 4x

3

1x

x2

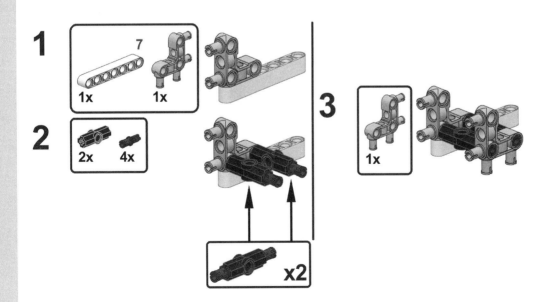

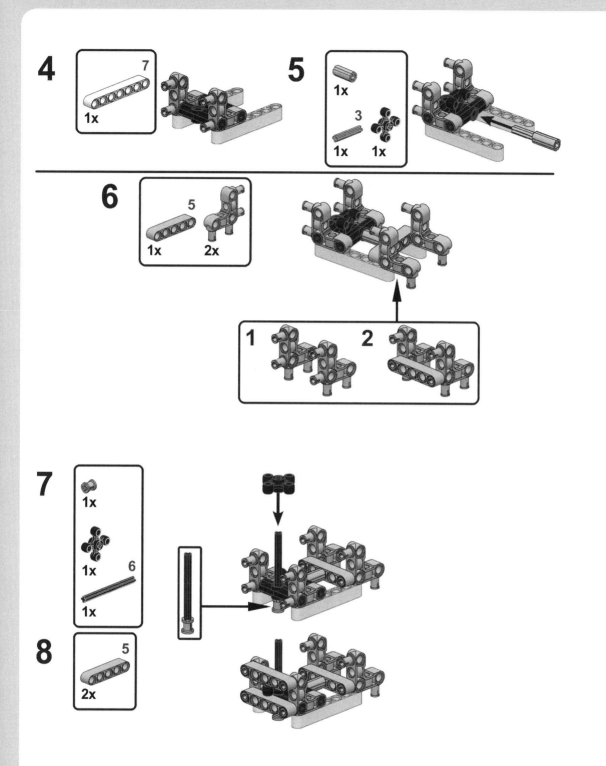

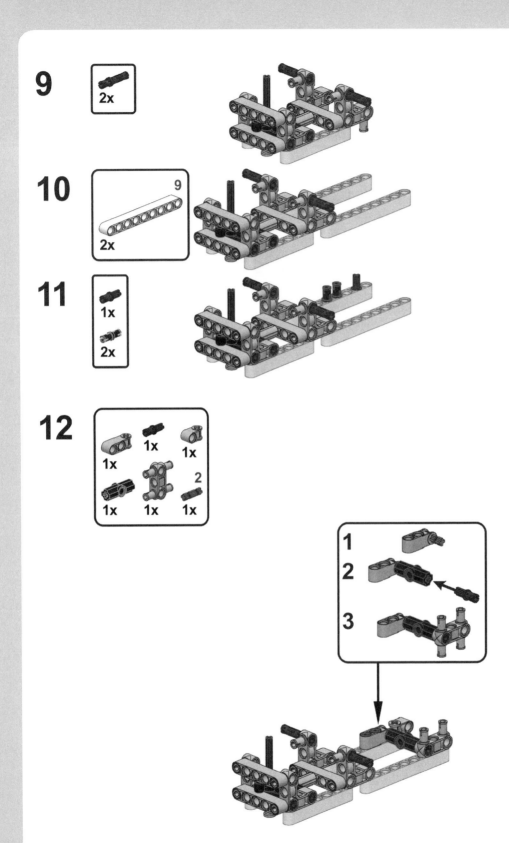

13

1x
12
1x

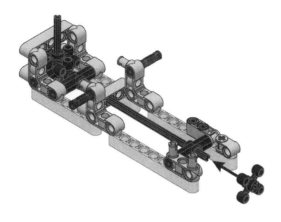

14

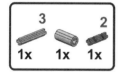

3
1x
1x
2
1x

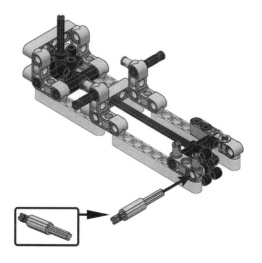

15

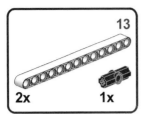

13

2x   1x

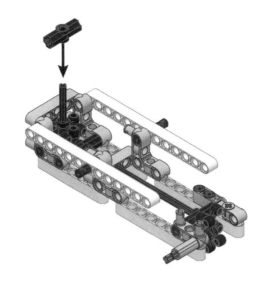

16

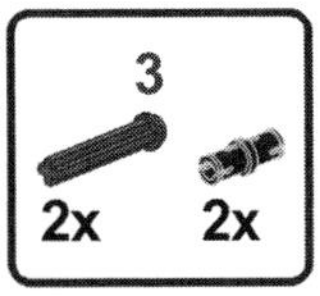

3

2x   2x

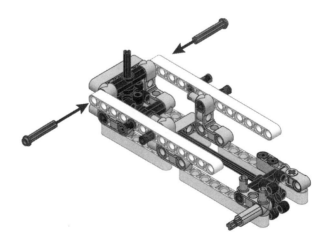

17

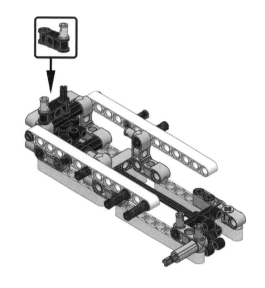

18

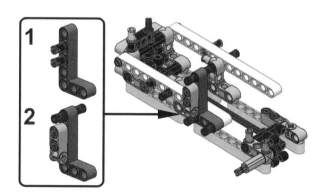

Return to the main assembly.

16

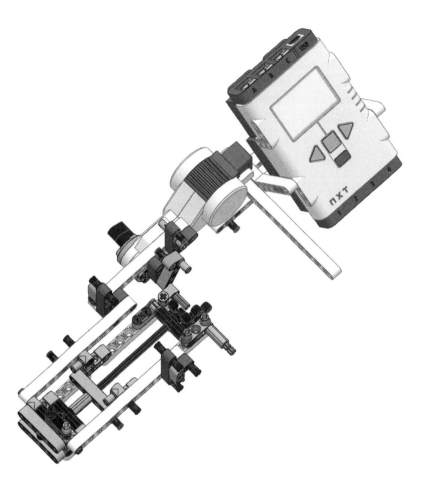

lever

1

13

1x **1x**

2

2x **5**

1x

3

5

2x

4

1x **13**

1x

5

2x

1x

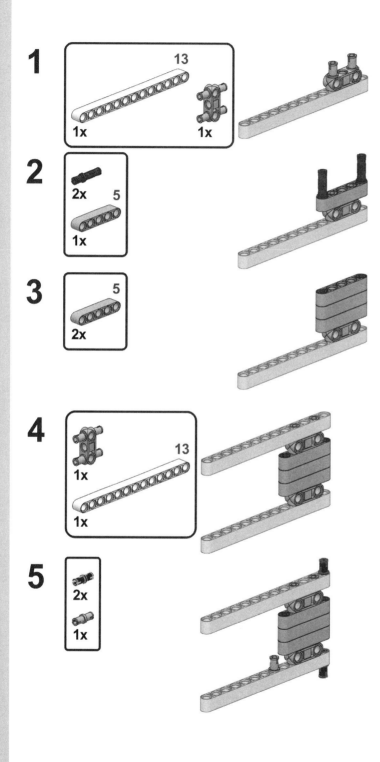

Return to the main assembly.

17

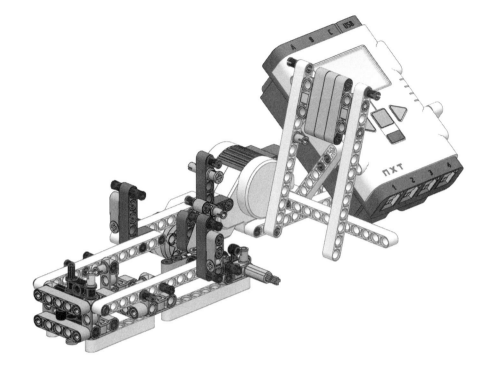

18

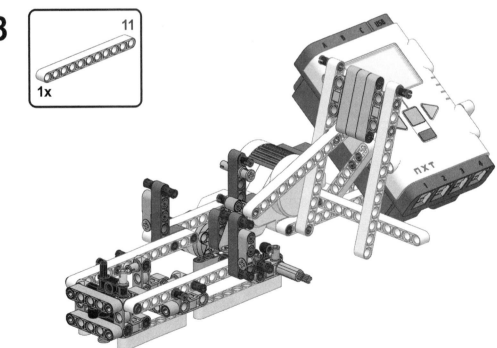

board motor

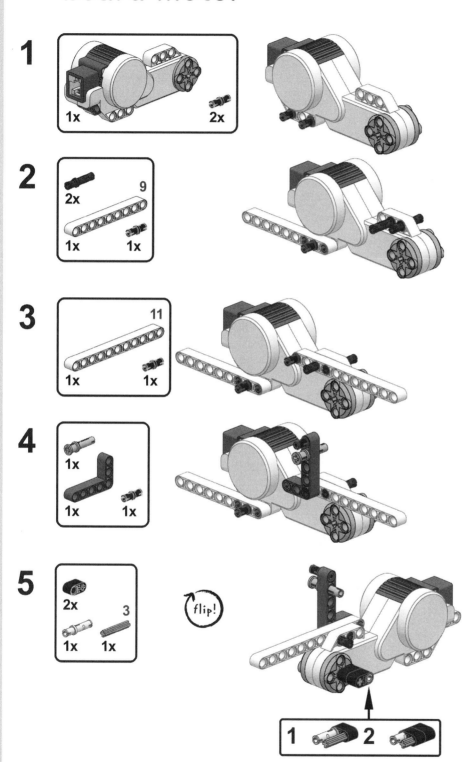

1
1x 2x

2
2x 9
1x 1x

3
11
1x 1x

4
1x
1x 1x

5
2x 3
1x 1x

flip!

1 2

Return to the main assembly.

19

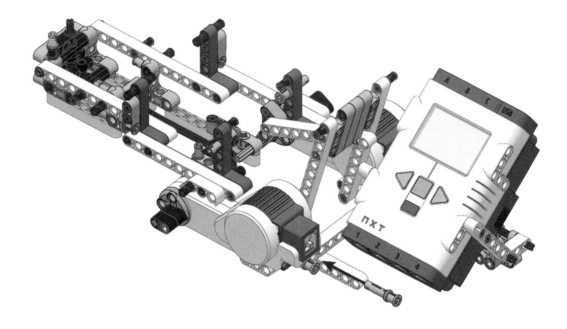

20

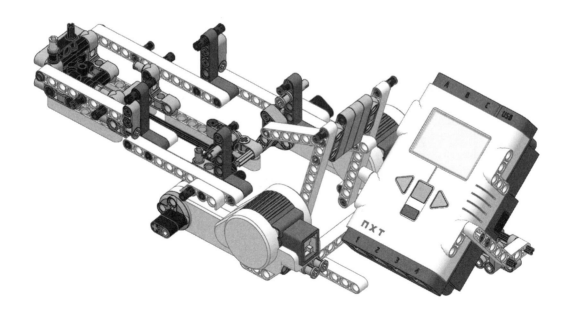

touch sensor

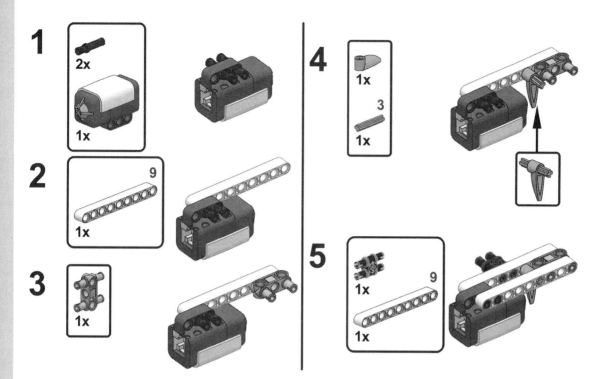

1 2x / 1x

2 9 / 1x

3 1x

4 1x / 3 / 1x

5 1x / 9 / 1x

Return to the main assembly.

21

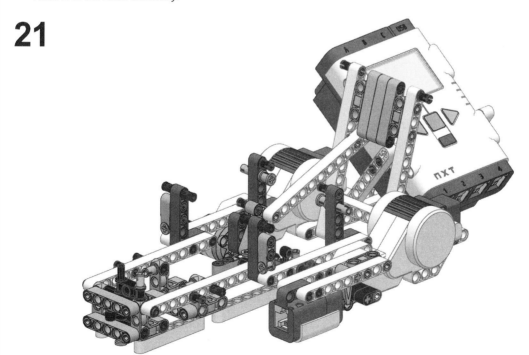

22

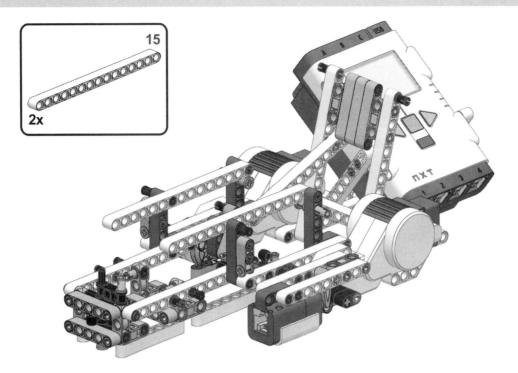

arm

1

1x
1x 2x

2

1x 7
1x

3

7
1x

4

3x
1x 5
1x

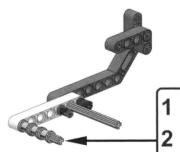

1
2

cursor guide

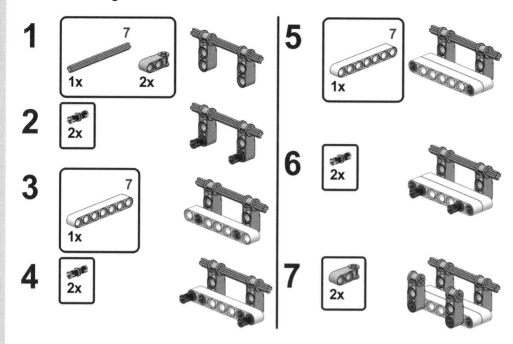

Return to the arm subassembly.

The cursor guide is shown aligned with the liftarms, but it will hang down loosely until you lock it in place later on.

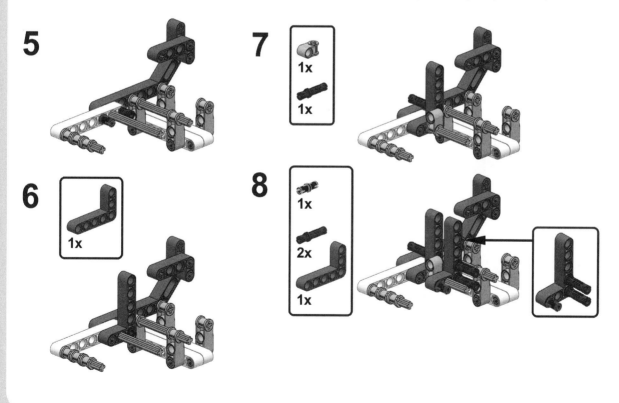

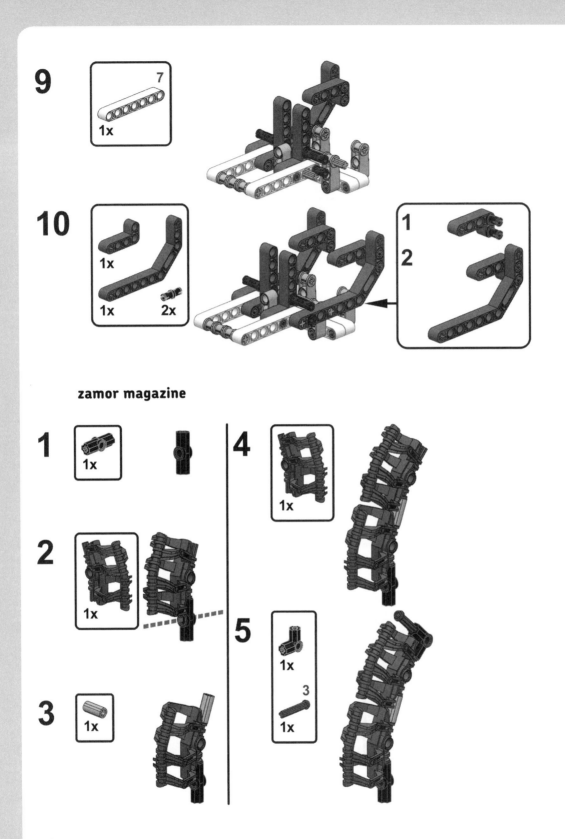

9

7
1x

10

1x

1x 2x

1

2

zamor magazine

1 1x

2 1x

3 1x

4 1x

5
1x

3
1x

6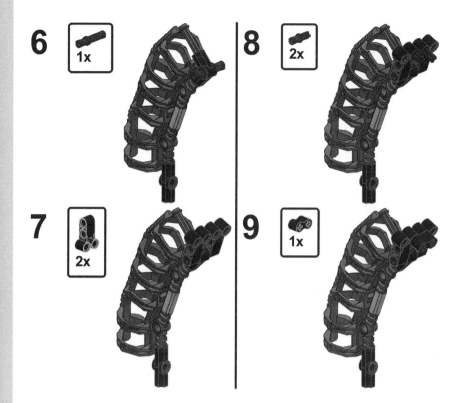

1x

8

2x

7

2x

9

1x

Return to the arm subassembly.

11

1x 5

1x

12

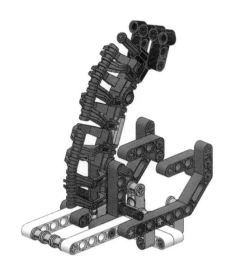

dropper motor

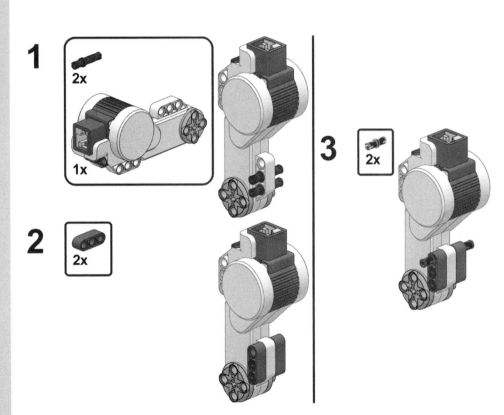

1 2x 1x

2 2x

3 2x

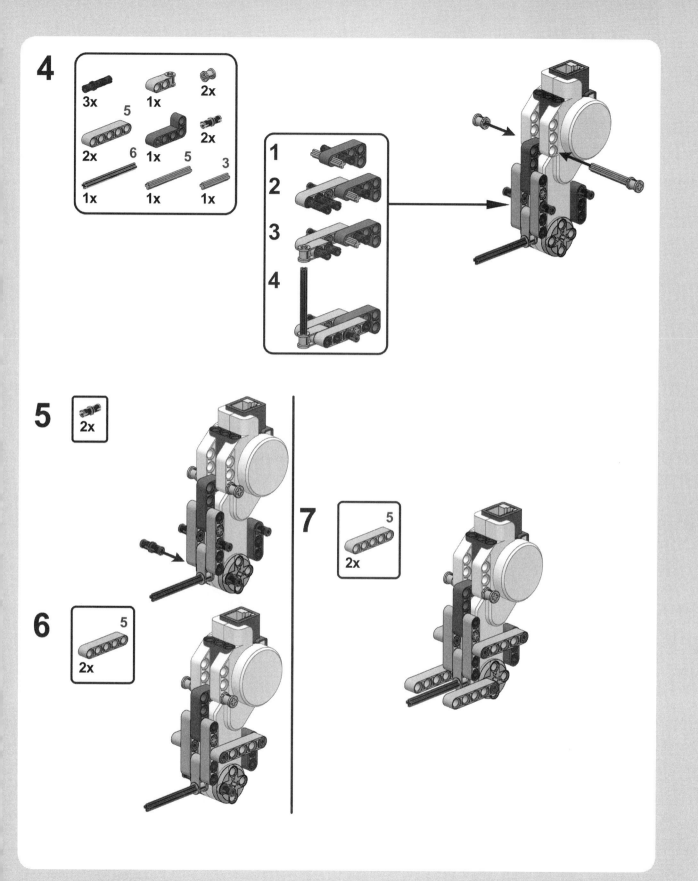

8

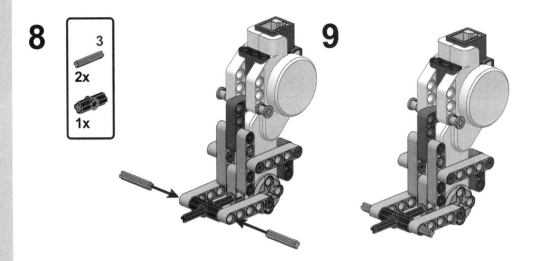

9

Return to the arm subassembly.

13

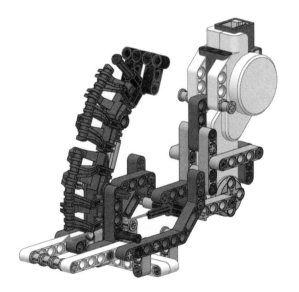

14

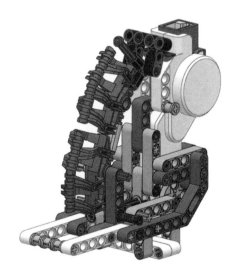

Open the dark gray liftarms slightly to allow the dropper motor subassembly to slide into place.

15

2x

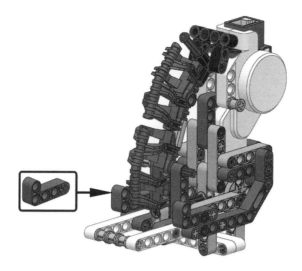

color sensor

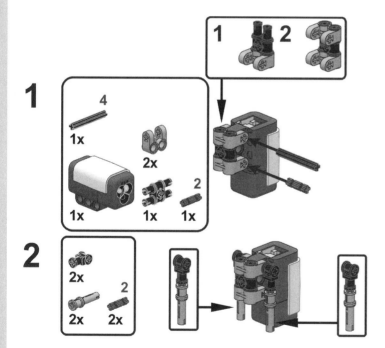

3

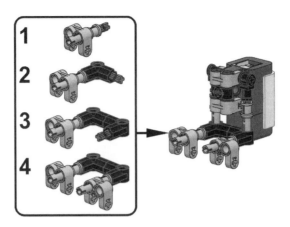

Return to the arm subassembly.

16

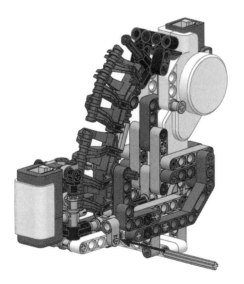

17

Return to the main assembly.

23

3

11

1x

1x

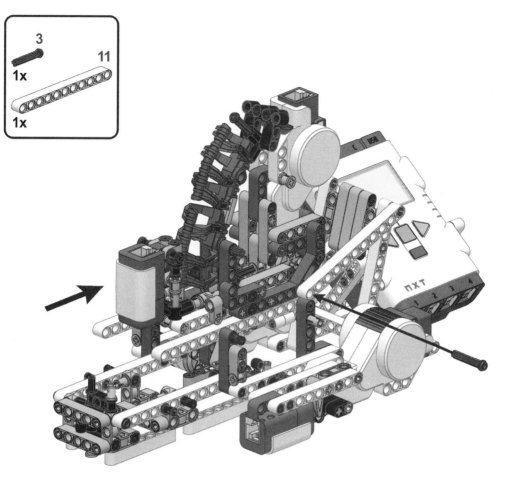

Be sure to fit the 3-long axle with stop into the slider subassembly to lock the cursor guide subassembly (shown on page 79) in place. The axle should fit in the cross hole of the light gray joiner.

24

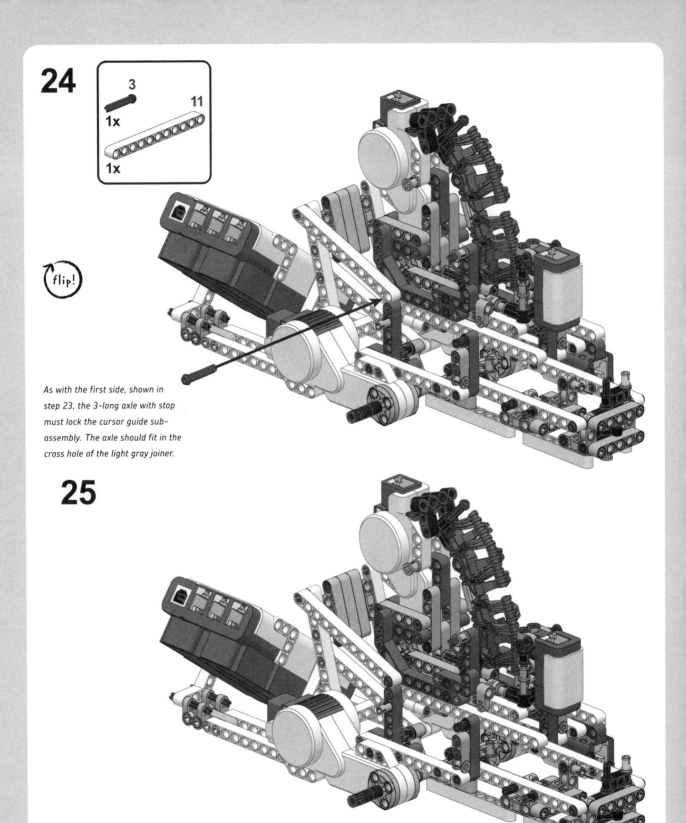

flip!

As with the first side, shown in
step 23, the 3-long axle with stop
must lock the cursor guide subassembly. The axle should fit in the
cross hole of the light gray joiner.

25

plate

1

2x **11**
1x **2x**

2

9
1x **2x**

3

7
1x **2x**

4

7
1x

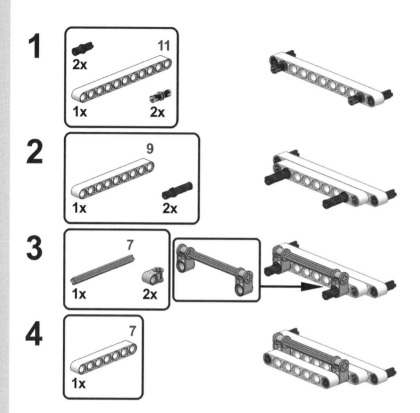

Return to the main assembly.

26

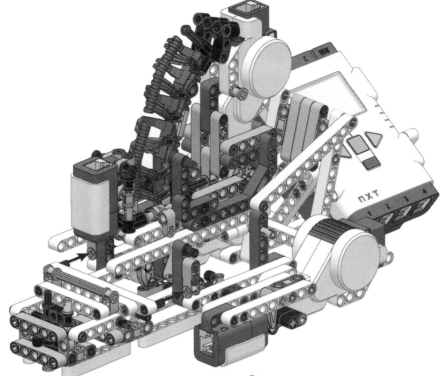

27

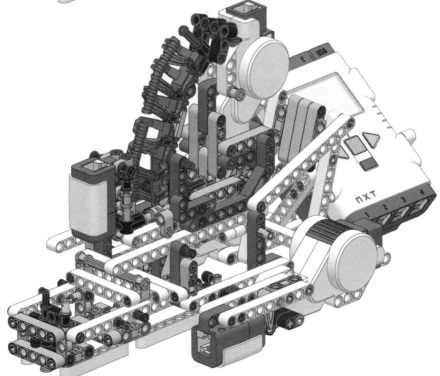

board

1 6x 2x

2 6x 7 2x

3 8x

4 7 7x

5 4x

6 4x

Return to the main assembly.

28

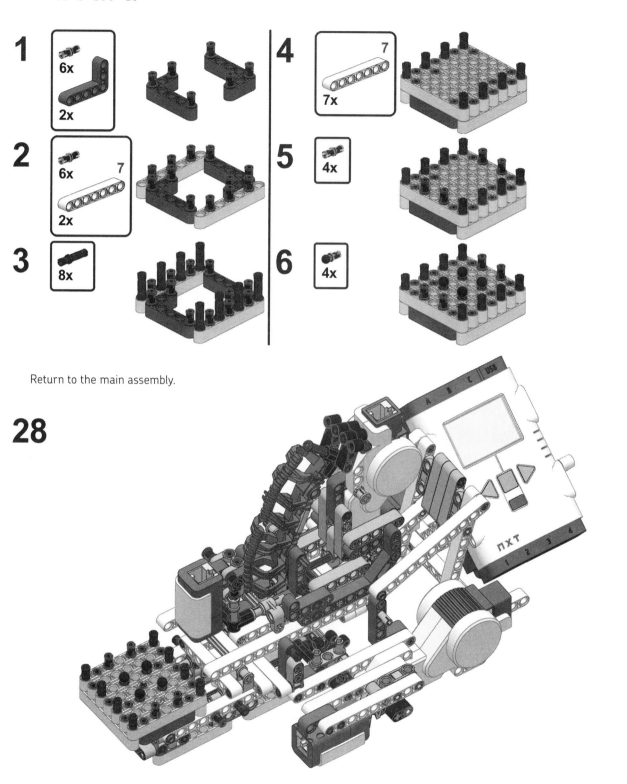

29

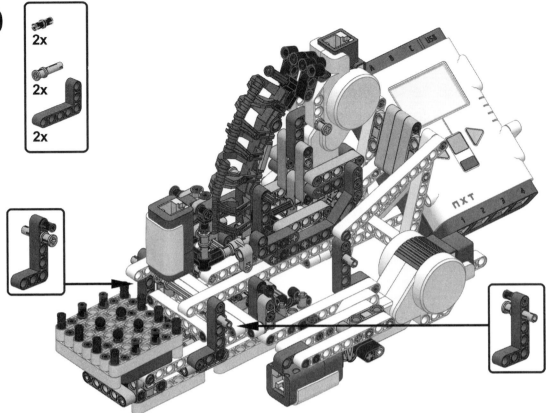

ultrasonic sensor

1

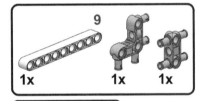

9
1x   1x   1x

2

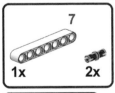

7
1x   2x

3

1x

Return to the main assembly.

30

31

20cm

1x

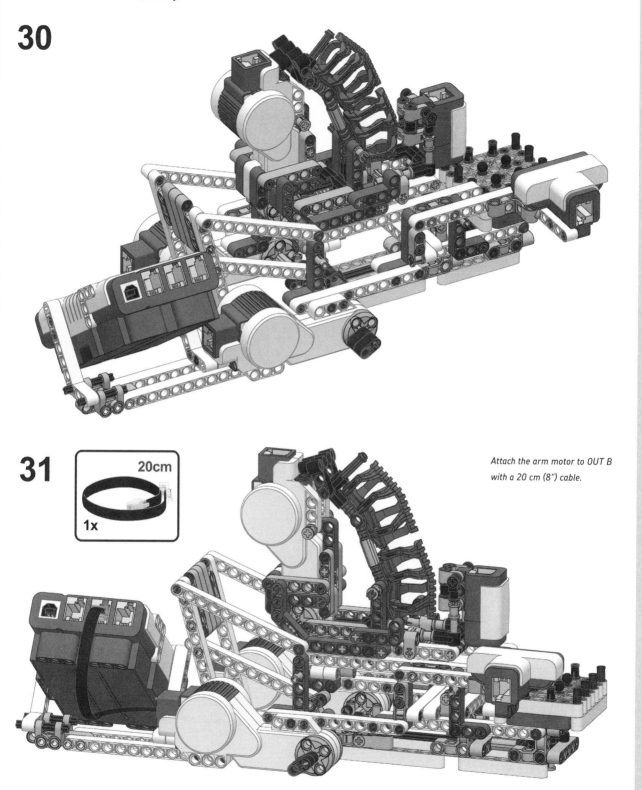

Attach the arm motor to OUT B with a 20 cm (8") cable.

32

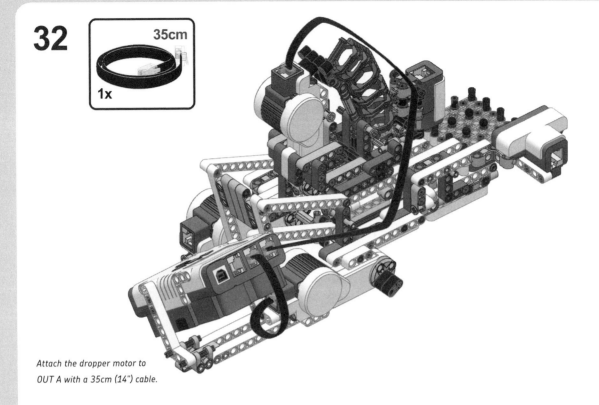

Attach the dropper motor to
OUT A with a 35cm (14") cable.

33

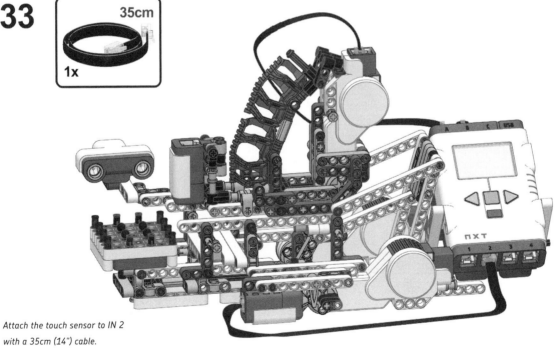

Attach the touch sensor to IN 2
with a 35cm (14") cable.

34

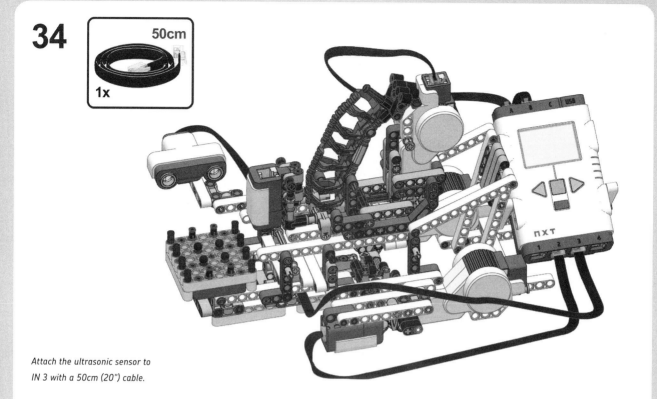

Attach the ultrasonic sensor to
IN 3 with a 50cm (20") cable.

35

Attach the light sensor to IN 1
with a 50cm (20") cable. Pass
the cable between the black pin
joiners above the Zamor magazine. The sliding arm should be
free to move (without the cable
obstructing it).

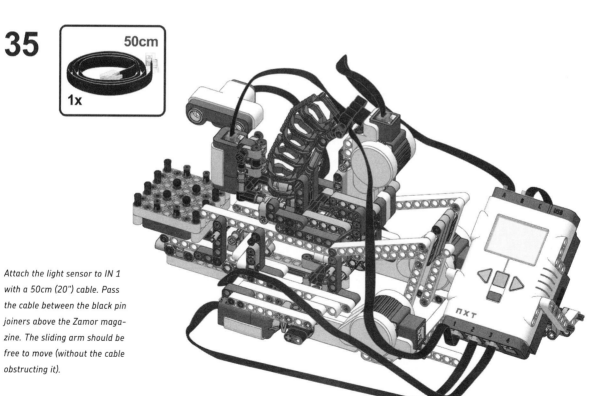

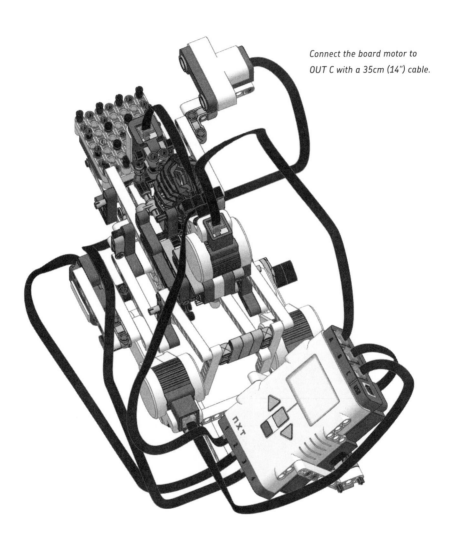

36

Connect the board motor to OUT C with a 35cm (14") cable.

37

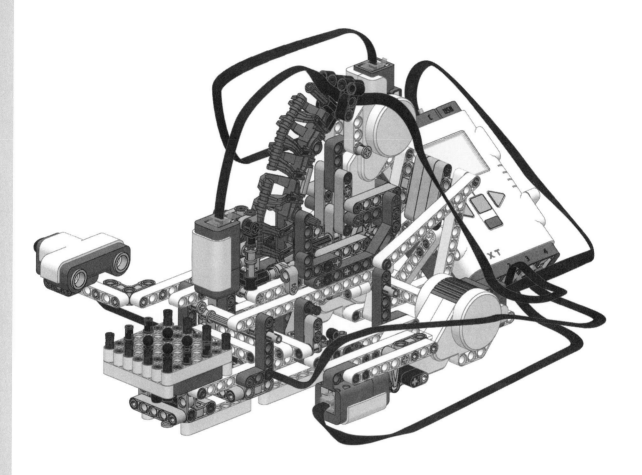

The TTT Tickler is ready to play!

5

the One-Armed Wonder: my new Rubik's Cube solver

In 2007, I began the LEGO Rubik Utopia (LRU) project, an artificial intelligence project, with the goal of developing a LEGO MINDSTORMS NXT robot that could automatically solve a Rubik's Cube. Within two years of its first appearance on the Web, the LRU project had millions of YouTube viewers, dozens of appearances on television around the world, and hundreds of requests for building instructions. And since then, many people have created their own LRU-inspired NXT robots.

In this chapter, you will be introduced to the One-Armed Wonder's curriculum vitae (who said robots can't have one?), and you'll meet its ancestors and offspring. You'll learn how to install the software needed for this robot to work on your computer, as well as how to use the robot and tweak its performance.

Like the TTT Tickler, the One-Armed Wonder comes in two versions, with building instructions for each NXT set in its own chapter (Chapters 6 and 7). Figures 5-1 and 5-2 show the two versions.

Figure 5-1: The One-Armed Wonder built with the NXT 8547 set

Figure 5-2: The One-Armed Wonder built with the NXT 8527 set

how the Rubik's Cube solver evolved

In April 2001, J.P. Brown built a robot that could solve a Rubik's Cube in about 10 minutes. For his robot, he used two RCX bricks, six motors, and a LEGO webcam. You can see his robot on his home page (*http://jpbrown.i8.com/cubesolver.html*). J.P.'s robot was groundbreaking, and people didn't attempt to improve on it for years.

In 2007, I developed some prototypes for solving a Rubik's Cube with the new NXT technology and came up with some complicated and bulky designs (you can see these at *http://robotics.benedettelli.com/LRU.htm*). I streamlined my initial designs in order to present the robot at the 2007 *Cirque Des Sciences* in Luxembourg. At that time, the robot looked like the one shown in Figure 5-3 and took about two and a half minutes to scan and solve the cube.

This version of the Rubik's Cube solver used three motors: one to push the cube, one to hold the two upper cube slices, and a third to rotate the base on which the cube was placed.

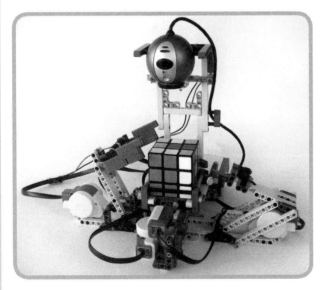

Figure 5-3: The original LRU robot, invented in 2007, in one of its later versions

In October 2007, the Rubik's Cube solver was the star of the inauguration of the Festival of Creativity in Florence. I was invited to demonstrate this amazing robot on many television shows, both in Italy and abroad. The YouTube videos of the robot reached millions of hits in a few months and have been shown at U.S. universities and on Japanese television. Not surprisingly, the requests for building instructions came in like a storm, and many other NXT robots inspired by mine began to appear.

Of the Rubik's Cube-solving robots, I think the most interesting is the Tilted Twister by Hans Andersson (*http://www.tiltedtwister.com/*). Andersson's robot can solve a cube in about six minutes by scanning the cube with a LEGO light sensor (without the use of a computer). It can be built with one NXT 8527 set, but it requires you to add modified color stickers to the cube in order for its light sensor to distinguish between the colors. This robot is interesting, but the LEGO sensor is quite picky about ambient light conditions, and the algorithm to find the solution for the cube takes a lot of time to run on the NXT.

Since the end of 2008 (and following many refinements), my robot has been able to solve a cube in *less than a minute*, including the time it takes to scan the cube (just eight seconds!). This speed places a lot of stress on the NXT servo motors, which undergo extreme torque and acceleration during the process. In fact, during its sparkling career, this robot has destroyed several parts and three motors: it has damaged a motor's gear train, cracked another motor's casing, and shattered a motor's orange shaft. To be sure that you don't destroy your own NXT set, I've designed a brand-new, gentler version of the robot for this book: the One-Armed Wonder.

The robot you'll build is a little slower than mine, but it will prove safer for your LEGO hardware. I've also optimized the resources needed so that the One-Armed Wonder uses just two motors: one to rotate the cube's support, and the other to drive a clever mechanism that tilts the cube, settles it on the support, and holds it in place as the top slices are rotated.

It's simpler to see this process in action than to explain it in words, and you can find a video of it on my website. In the future, I may add an arm and a third motor to scan the cube with a LEGO color sensor, like the Tilted Twister had.

requirements for the One-Armed Wonder

The One-Armed Wonder needs to be connected to a computer via USB. Once connected, and with the correct programs running on your robot, you'll run the CubeSolver program on your computer to take snapshots of the cube with a webcam, map the *facelets* (the small colored squares on the cube), compute a solution, and send the robot the sequence of movements need to solve the cube.

You should be able to run the CubeSolver program on any computer running Windows 2000 or higher. The instructions in the next section, "Installing the Software," will tell you how to set up the software on your computer.

NOTE If you have any problems running this software, please visit *http://tr.benedettelli.com/* for help.

You'll also need to install a USB webcam on your robot. In my photos, I've used an older webcam called the Logitech QuickCam Chat, which is ideal for mounting with LEGO parts because it has a hole in the bottom and two holes on the sides. (It also looks kind of cool.) If you can't find this particular camera, look for a similar one, like the Logitech Webcam C200, C250, or C500. These cameras do not have the same holes as the QuickCam Chat, but the holes in the bottom and on the side (accessible once you remove the rubber stopper) should be enough to lock it in place. You can also use the LEGO Logitech Camera, included in the LEGO MINDSTORMS Vision Command 9731 set, released in 2000.

NOTE You should be able to use any brand of camera, but be sure to get one that is compatible with your operating system! For example, my Logitech QuickCam Chat won't work with Windows Vista or 7, and the same is true for the LEGO MINDSTORMS Vision Command camera. Also, be sure to disable the automatic gain control in the camera's software (if available) to make the color detection more reliable.

installing the software

In order to control your robot, you'll need to download and run certain programs on both your robot and your computer. Visit *http://tr.benedettelli.com/* or *http://www.nostarch.com/* for links to the software you'll need, which is free.

NOTE These programs may change over time in response to reader feedback. Visit my website for updates.

the CubeSolver program

To begin, download the CubeSolver program from my site, save it to your desktop, and extract it. You should end up with a folder named *CubeSolver*. Double-click the folder to open it and look for the *CubeSolver.exe* file. (You'll see some additional files as well. Just leave them there.)

Before running the CubeSolver program, be sure that you have the following software installed on your computer.

NXT driver 1.02

If you have not already installed the LEGO MINDSTORMS NXT driver update (required for communication with the NXT brick), download the MINDSTORMS NXT Driver 1.02 for Windows from *http://mindstorms.lego.com/support/files/* and install it on your computer.

Microsoft DirectX 9

Make sure that your computer has DirectX version 9 or higher as follows:

1. Click **Start ▶ Run**.

2. Enter **dxdiag** and click **OK**.

3. The DirectX Diagnostic Tool should open. On the System tab, make sure that the version is 9 or higher on the DirectX Version line.

4. Click **Exit** once you've finished.

In Windows Vista or 7, click the Windows icon, then enter **dxdiag** in the search box and press ENTER.

using the CubeSolver program

This section describes how to start the CubeSolver program and use its controls. If you haven't already done so, download and extract the *TRfiles.zip* archive from *http://tr.benedettelli .com/* or *http://www.nostarch.com/*. There should be two programs inside the archive:

LRU09_1.rxe   Use this program if your NXT is running firmware version 1.05.

LRU09_2.rxe   Use this program if you are running firmware version 1.28.

NOTE For instructions on downloading files to the NXT, see "Quick Start Guide for Windows Users" on page 4.

To run the CubeSolver program, double-click *CubeSolver.exe*. This program file should appear with a Rubik's Cube icon. (This icon represents the first Rubik's Cube configuration that the early prototype of LEGO Rubik Utopia was able to solve.) Figure 5-4 shows the program's graphical interface.

NOTE If for some reason the CubeSolver won't run, make sure that you've installed the NXT driver update and that DirectX is up to date. If you need further assistance, visit the forums at *http://tr.benedettelli.com/*.

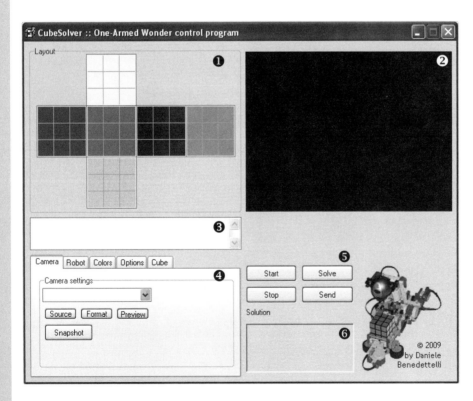

Figure 5-4: The CubeSolver program elements

the program controls

In this section, I'll cover the CubeSolver's interface and controls in detail, as shown in Figure 5-4. Some of these elements adjust advanced functions or settings that you may not need to use immediately. If you are impatient to see your robot in action, skip this section and go to the next one!

the cube layout panel

The Cube Layout panel (labeled ❶ in Figure 5-4) shows the cube as if it were laid out flat, like taking a box apart and laying its faces on a table. Figure 5-5 shows how the flattened faces on the screen map to the faces on the actual cube.

You can change the color of every *facelet* (the face of each small square on a cube face) while in Edit Mode by left-clicking on that facelet. To change the color applied to a facelet, left-click one of the six color buttons in the palette, or right-click the facelet to sample its color.

When using the webcam you can change a whole face (all nine facelets) by dragging the Video Window (labeled ❷ in Figure 5-4) to one of the faces on the Cube Layout panel. The colors should be assigned in the order in which they are displayed in the Video Window.

the video window

The Video Window (labeled ❷ in Figure 5-4) shows the video as captured by the webcam, with nine small squares superimposed over the video image. These squares are there to help you to align the camera to look directly onto the top face of the cube. The color of each square shows the program's guess at the color of that particular facelet. Move the Spread slider to adjust the spread of these squares, adjusting it until the sample colors are about in the middle of each facelet.

the log

The Log (labeled ❸ in Figure 5-4) reports information or warnings about events that have occurred with the program or your robot. For example, it records the successful generation of the solver algorithm tables and warns you about the absence of a webcam or the NXT. Double-click the log to clear it.

the control panel

The CubeSolver's Control Panel (labeled ❹ in Figure 5-4) has five tabs: Camera, Robot, Colors, Options, and Cube.

the camera control tab

The Camera Control tab, shown in Figure 5-6, has four buttons and a drop-down menu listing all available video-capture devices on your system. Select your camera from this list.

* The Source and Format buttons open dialogs that allow you to configure your camera. (We'll talk more about these later in this chapter.)
* The Preview button allows you to toggle the webcam's video preview and the real-time color recognition. When the camera is not in preview mode the Spread slider is hidden.
* Click the Snapshot button to save an image called *capture.jpg* to the same folder as the CubeSolver executable.

Video Preview

Figure 5-5: The cube layout

Figure 5-6: The Camera Control tab

the robot control tab

The Robot Control tab, shown in Figure 5-7, contains three buttons: Get Info gets basic information from the NXT brick, and Demo 1 and Demo 2 start demo sequences that are useful for testing the robot. Clicking **Demo 1** tells the robot to change the cube orientation and display certain information on the NXT screen that will be useful for debugging; clicking **Demo 2** tells the robot to execute a particular sequence of moves in order to bring a solved cube into a particular configuration.

When the Beep checkbox is selected, the NXT will play a short beep to acknowledge that it received the command you sent whenever you click a button in the CubeSolver software.

Figure 5-7: The Robot Control tab

the colors tab

Use the Colors tab (shown in Figure 5-8) to change the parameters of the color recognition algorithm. This algorithm guesses the color of each of the nine visible facelets by analyzing the snapshot of the cube face captured by the camera. The color recognition algorithm is very good at distinguishing blue and green, but you'll probably need to adjust it a little for it to correctly recognize red, orange, and yellow. Here's how to use each parameter:

* Increase the White parameter to cause darker colors to be seen as white. Lowering this value tells the algorithm that only very bright areas should be considered white.

* The Red parameter sets the upper threshold for the color red. Tweak this parameter to allow the algorithm to correctly distinguish red from orange. If you set this value too low, the program might mistake orange for red.

* The Orange parameter is the upper threshold for the color orange. Tweak this parameter to allow the algorithm to correctly distinguish orange from yellow.

* The Yellow parameter sets the upper threshold for the color yellow. This is not a particularly sensitive parameter, but you may find it useful in certain cases.

When the camera preview is enabled, the changes you make in the Colors tab take effect immediately. To revert to the default Color settings, click the **Reset** button and answer **Yes** when asked to confirm.

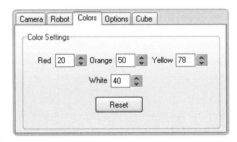

Figure 5-8: The Colors tab

the options tab

As you can see in Figure 5-9, the Options tab allows you to change the order in which the faces of the cube are scanned and their orientation with respect to the camera.

WARNING The default settings are designed to work with the LRU09_x.rxe program and *must not be changed* (unless you want to develop your own NXT software for the robot). If you accidentally change these settings, click the Reset button to revert to the defaults and answer Yes when asked to confirm.

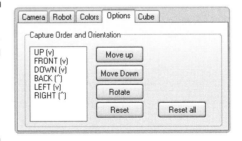

Figure 5-9: The Options tab

To change the order, select a face name in the list and then click either the **Move Up** or **Move Down** buttons. To change the orientation of a face, select it, click one of the Orientation buttons as described below, then press **Rotate**. (The symbol inside the brackets will change to reflect that face's rotation.) Specifically:

(^)   Copies the nine facelets in the Video Window to the face in the Cube Layout panel (section ❶ in Figure 5-4) without rotation

(>)   Copies the nine facelets, rotating them 90 degrees clockwise

(v)   Copies the nine facelets, rotating them 180 degrees

(<)   Copies the nine facelets, rotating them 90 degrees counterclockwise

The Reset All button deletes all configuration files used by the program and restores its default settings. You shouldn't need to use this button very often.

the cube tab

The Cube tab, shown in Figure 5-10, allows you to edit the virtual cube shown in the Cube Layout panel. This tab has the following components:

* In the Manual Input panel, you can rotate the cube faces clockwise by left-clicking, or counterclockwise by right-clicking on the buttons that represent the faces.
* The Edit Mode button changes the program to Edit Mode, which allows you to edit the cube colors as explained earlier. While in Edit Mode, the text in the Edit button changes to say Confirm. To commit the changes you've made to the cube, click the **Confirm** button and the program will exit Edit Mode.

NOTE When you exit Edit Mode and the cube configuration is not acceptable (for example, if there are more than nine facelets of each color), an error message should appear asking whether you want to continue editing or reset the cube layout. If the cube is not consistent, the solver algorithm won't work.

* The Shuffle button shuffles the cube, applying a number of random moves to it as determined by the drop-down list.
* The Reset button resets the cube and the program interface.

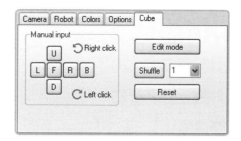

Figure 5-10: The Cube tab

high-level control buttons

Four buttons on the lower-right side of the main CubeSolver program window (labeled ❺ in Figure 5-4) control the robot's high-level functions, as follows:

Start   To start the complete scanning and solving procedure, press the **Start** button. If all goes as planned, your robot should solve the Rubik's Cube! (If something does go wrong, see "Troubleshooting" on page 107.)

Stop   To stop the robot at any time, press the **Stop** button.

Solve   If you are working offline (using the program to solve a cube edited by hand, without the aid of the webcam and robot), you can find the solution for the cube by clicking the **Solve** button. The solution will be displayed in the Solution panel (labeled ❻ in Figure 5-4).

Send   Click the **Send** button to tell the robot to begin solving the cube, executing the move sequence shown in the Solution panel (discussed below). Use the Send button when you have already mapped the cube and you want to skip the robot's initial cube-scanning sequence and just solve it.

the solution panel

The Solution panel (labeled ❻ in Figure 5-4) shows the sequence of moves necessary to solve the cube as last calculated by the program. For example, a sequence could be U D' R L2 F'. You will learn how to interpret the notation used in this sequence in "A Little Rubik's Cube Notation" on page 107 so that you can use it to solve the Rubik's Cube by hand.

solving the cube has never been so easy!

Now I will show you how to solve your cube with the One-Armed Wonder. The solution is only a few clicks away! The entire cube-solving process should run automatically, but things sometimes go awry. (If they do, please see "Trouble-shooting" on page 107 or visit *http://tr.benedettelli.com/* for further assistance.)

NOTE Be sure to lubricate the cube before using it with the robot, as discussed in "Robot Fails to Rotate the Cube" on page 107, in order to make the faces easier to turn.

Follow these steps to solve the cube:

1. Turn on the NXT and start the LRU09_*x* program that matches your version of the NXT. *LRU09_1.rxe* is designed to run on the original NXT; *LRU09_2.rxe* is for the NXT 2.0. The robot should reset its moving parts.

2. Connect the webcam and the NXT to your computer, then launch the CubeSolver program.

3. Wait for the program to generate the solution tables needed by its solver algorithm. While the program works, the Solve and Start buttons are disabled. When the program finishes generating the tables, the log should report `Solver tables generated`, and the Solve and Start buttons should be enabled.

4. The CubeSolver program should automatically connect to the first available video capture device, displaying the device's stream in the Video Window. If you have multiple cameras attached to your computer and the wrong video is displayed, select the **Camera** tab and choose your camera from the drop-down list.

5. Click the **Preview** toggle button to switch the webcam preview on and off. Preview should be active when solving the cube.

6. Put the cube on the robot's rotating base under the webcam, then adjust the camera position so that the colored squares on the video screen line up with the cube's facelets. To adjust the spread of the sampling area, move the **Spread** slider as discussed on page 103. If the colors are not properly recognized, tweak the camera settings as discussed on page 104.

7. Click the **Format** button to open the camera's format settings. For best performance, choose a 320 × 240 resolution and RGB24 pixel format. Also, be sure to turn off any image mirroring—the Video Window should display exactly what the camera is seeing, without any mirroring or rotation (as shown in Figure 5-5). Click **OK** to accept the changes and close the dialog.

8. Click the **Source** button to display the camera settings dialog. (This dialog will look different for every webcam.)

9. Click the **Default** button to reset the camera's settings. Next, disable automatic gain control and adjust the camera's other settings until the colors in the Video Window look natural.

10. Click **OK** to accept your changes and close the dialog.

At this point, make sure that the camera is recognizing the colors correctly, rotate the cube, and fine-tune the settings as necessary. If everything looks okay, continue with these steps:

1. Select the **Robot** tab, then press the **Get Info** button. If the NXT is properly connected, it will beep and the log will display its name, the running program, and the battery level. If it is not properly connected, the log will show `NXT is off or not connected`. (If you see this message, make sure that the NXT is turned on and connected to your computer and that the drivers are properly installed.)

2. Shuffle the cube, making sure that the faces are flat and that the cube slices are aligned in relation to each other (to reduce the risk of the cube getting stuck).

3. Place the cube on the rotating base.

4. Click **Start** and prepare for the wonder!

NOTE The first time you use the CubeSolver program, start the LRU09_*x* program on the NXT that matches your version of the kit (where *x* is 1 or 2, depending on the version of your NXT set). The next time you run CubeSolver, the correct NXT program should start automatically.

If everything worked perfectly, congratulations! If something went wrong, read the next section for troubleshooting tips.

troubleshooting

Here are solutions to some common problems that you might encounter with your cube-solving robot. For additional troubleshooting suggestions, as well as updated programs and instructions that might help with a particular problem that you're encountering, please visit *http://tr.benedettelli.com/*.

NOTE You can always stop the robot by clicking the Stop button in the CubeSolver software or by pressing the dark gray button on the NXT brick.

robot fails to reset its arm

If the robot fails to reset its arm, the most probable cause is that the touch sensor is not properly closing when the arm is in 0 position. To fix this, try moving the touch sensor closer to the camshaft. To see if the touch sensor closes when the arm is completely retracted, use the NXT on-brick View menu, selecting Touch Sensor on port IN 1. When the arm is in 0 position, the sensor should read **1**.

robot fails to rotate the cube

If your robot fails to rotate the cube properly or at all, try lubricating the inside of your cube before using it with your robot by spraying silicone lubricant between the faces. Once you've sprayed the lubricant, rotate the cube for a while as the lubricant dries or it will end up acting more like a glue than a lubricant. (See *http://lar5.com/cube/speed.html* and *http://jpbrown.i8.com/cubesolver.html* for details about the lubrication procedure used by speed-cubers.) Also, be sure that the cube's stickers are securely attached to the cube and that their corners are not crumpled. Poorly attached stickers may be a cause of failure.

Another option is to try adjusting the robot hand's grasp. The hand's grip on the cube should be firm, but there should be only a minimal amount of friction between the hand and the faces of the cube. Also, especially with older cubes, the stickers may stick to the robot's hand. If that happens, try adjusting the hand's grasp by slightly spreading the hand's fingers.

After using your robot for a while, the white rubber bands used by the tilting mechanism may become slack. If this happens, replace them.

a color is incorrectly identified

When a color cannot be identified, the cube-scanning sequence stops and a pop-up message asks you to correct the facelet colors. You are now in Edit Mode. The Cube tab should be selected and the Edit toggle button should have changed to Confirm.

To correct the colors use one of the following methods:

* Select the currently assignable color by clicking one of the six colors on the color palette, or right-click a facelet to sample its color and make that color the current one.
* Click a facelet to assign the current color to it.

If the colors on the cube become consistent while in Edit Mode, the log should report the message Cube is OK! Once you've finished adjusting the colors, click **Confirm** to continue. If there are still problems with the cube, another error message will pop up and you will be asked whether to continue editing or cancel. Once you've fixed any errors, the program should resume finding the cube solution and send it to the robot.

Certain ambient light conditions (such as the warm, yellowish light from a halogen bulb) may also prevent the robot from recognizing the colors correctly. Try moving the robot so that the light illuminates the cube differently, and play with the camera settings.

a little Rubik's Cube notation

The standard Rubik's Cube notation was invented by David Singmaster and is known as *Singmaster notation*. The color of a Rubik's Cube face is determined by the color of the central facelet on that face. Table 5-1 should help you to interpret the move sequences in Singmaster notation.

table 5-1: singmaster notation

Key	Meaning
R	Right face
L	Left face
U	Up face
D	Down face
F	Front face
B	Back face
'	Suffix meaning reverse move (counterclockwise)
2	Suffix meaning double move (180 degrees)

Here's an example of how to use this notation:

```
R L' B2 D
```

This sequence translates as follows:

1. Rotate the right face clockwise.

2. Rotate the left face counterclockwise.

3. Rotate the back face by 180 degrees.

4. Rotate the down face clockwise.

When executing this move sequence by hand, be sure that the cube is always kept in the same orientation or the face names will change.

NOTE In this notation, a face's name alone (such as R) means rotate that face clockwise by 90 degrees; a face's name followed by an apostrophe (such as R') means rotate that face counterclockwise 90 degrees. If a 2 follows a face's name (such as R2), rotate that face 180 degrees (two full turns). The clockwise and counterclockwise rotations assume that you are looking directly at the cube face you are turning.

summary

This chapter began with the history of the Rubik's Cube solver robot that you'll build in the next chapters. Then you learned how to set up the CubeSolver software on a computer running Windows, how to use the CubeSolver program, and how to troubleshoot common problems with your robot. Finally, you saw some of the standard notations for Rubik's Cube move sequences.

6

building the One-Armed Wonder with the original NXT set

In this chapter you will build the One-Armed Wonder using the original NXT retail set 8527. In addition to the parts in your set, you will need a USB webcam (see Chapters 1 and 5 for details), two LEGO white rubber bands (or equivalent ones, about 1.5cm (0.59") in diameter, with the same elastic tension), and a 65mm × 55mm (2.56" × 2.17") cardboard or plastic slippery sheet to attach to the rotating base.

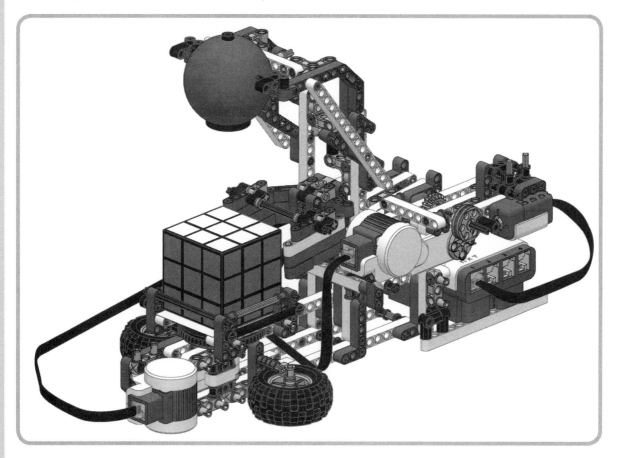

bill of materials

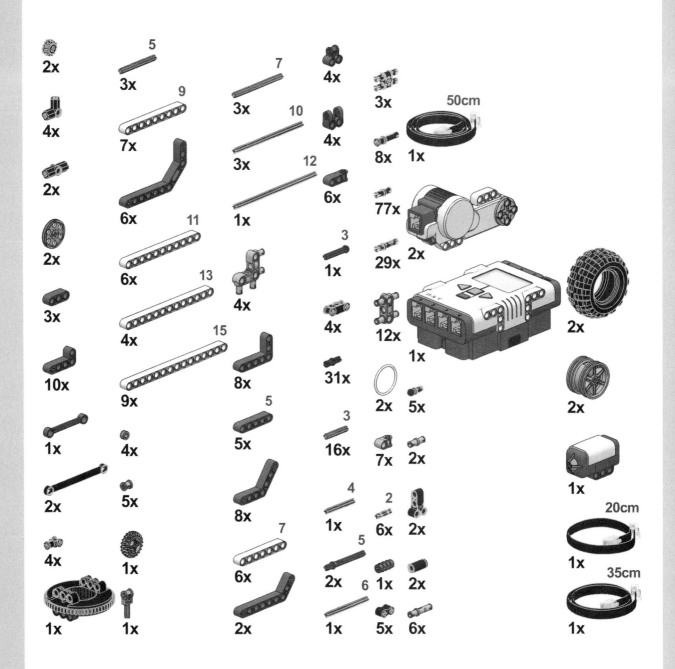

main assembly

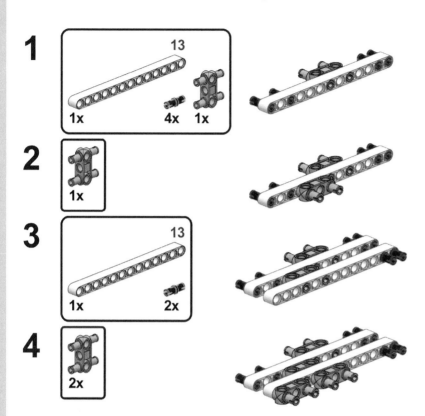

1

13
1x   4x   1x

2

1x

3

13
1x   2x

4

2x

base motor

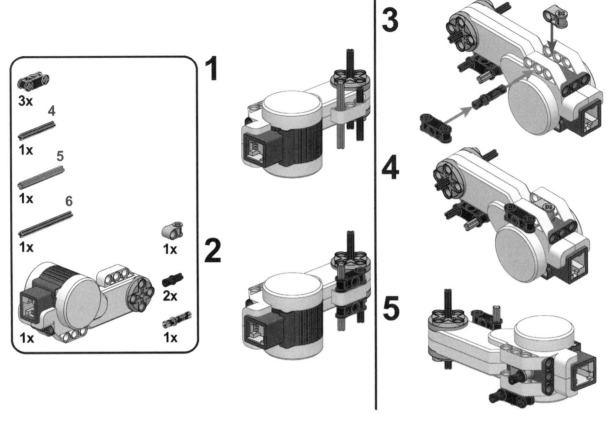

Return to the main assembly.

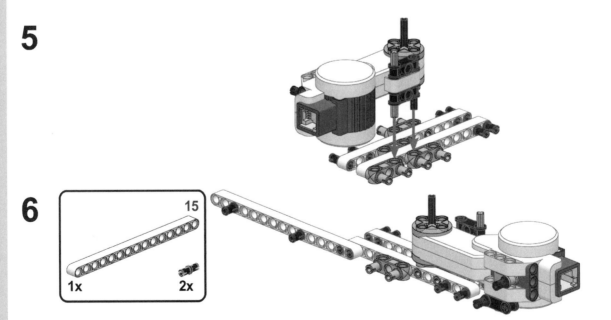

base top layer

1

11

1x 3x

2

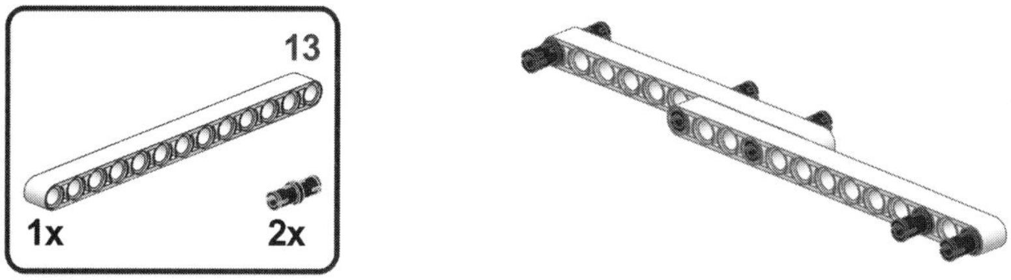

13

1x 2x

3

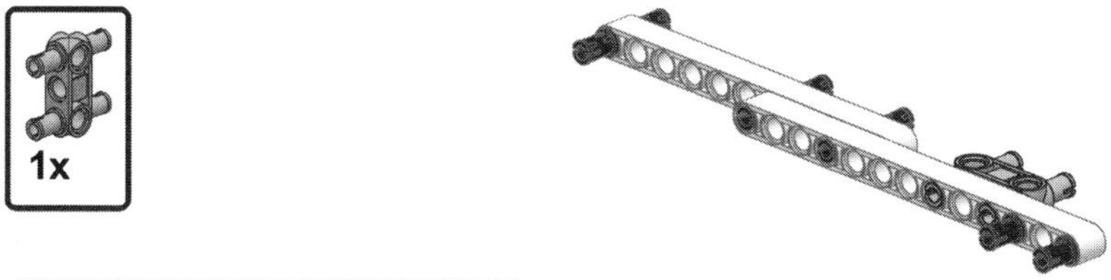

1x

4

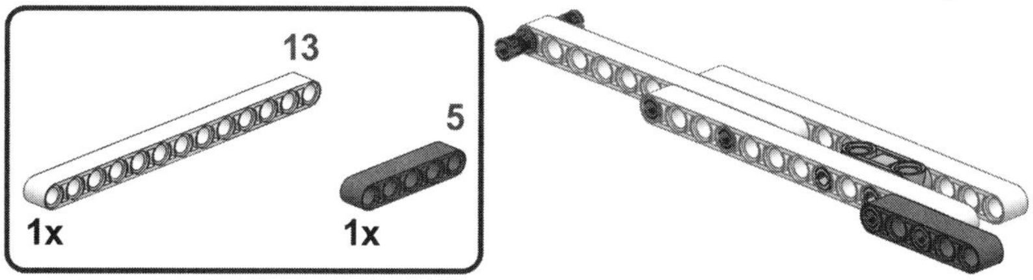

13

5

1x 1x

5

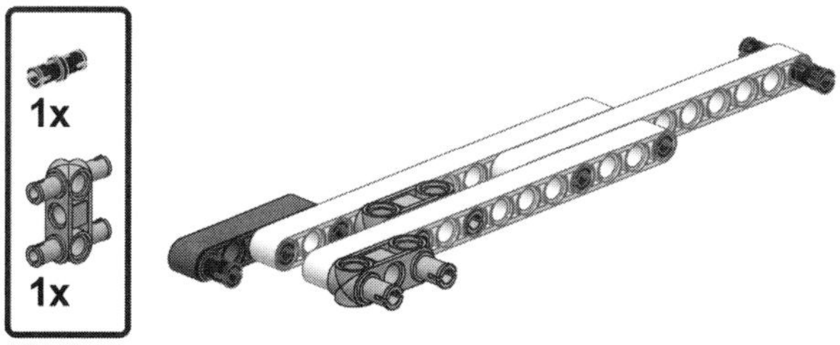

1x

1x

6

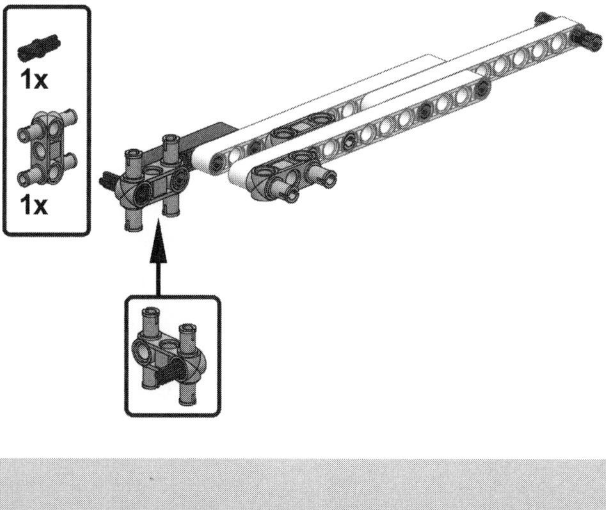

1x

1x

Return to the main assembly.

7

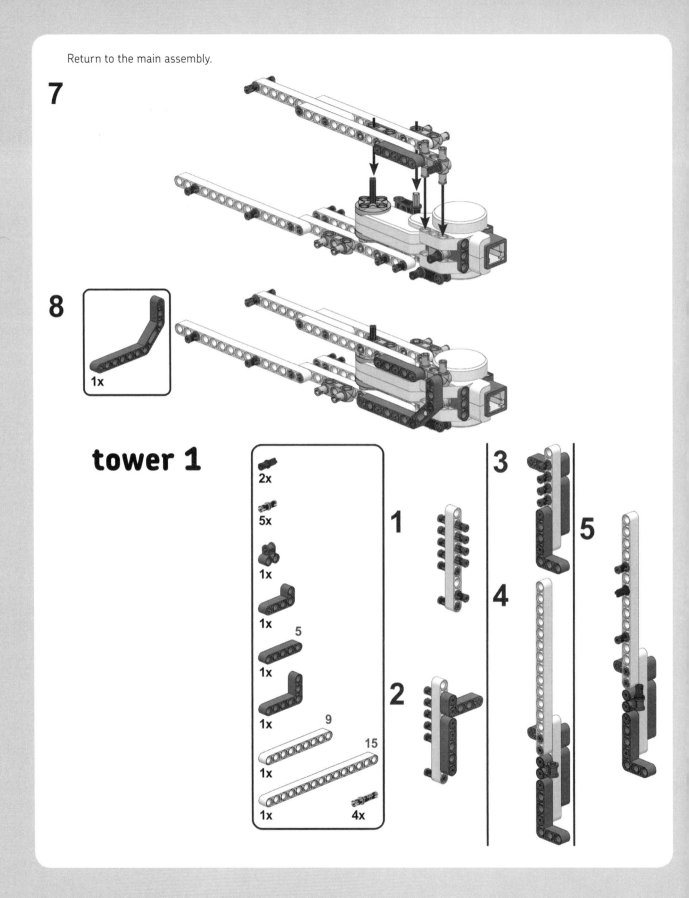

8

1x

tower 1

2x

5x

1x

1x

1x 5

1x

1x 9

1x 15

1x 4x

1

2

3

4

5

Return to the main assembly.

9

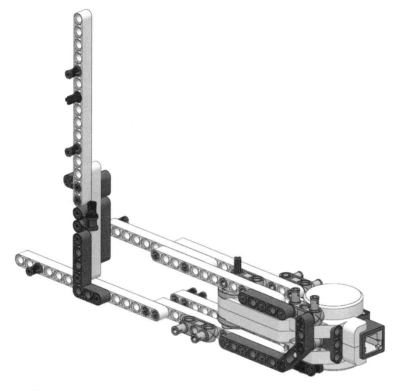

tower 2

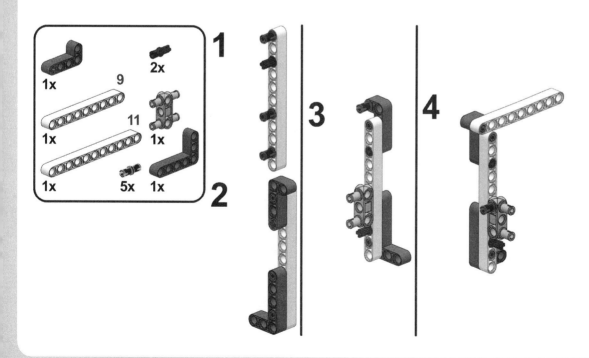

Return to the main assembly.

10

11

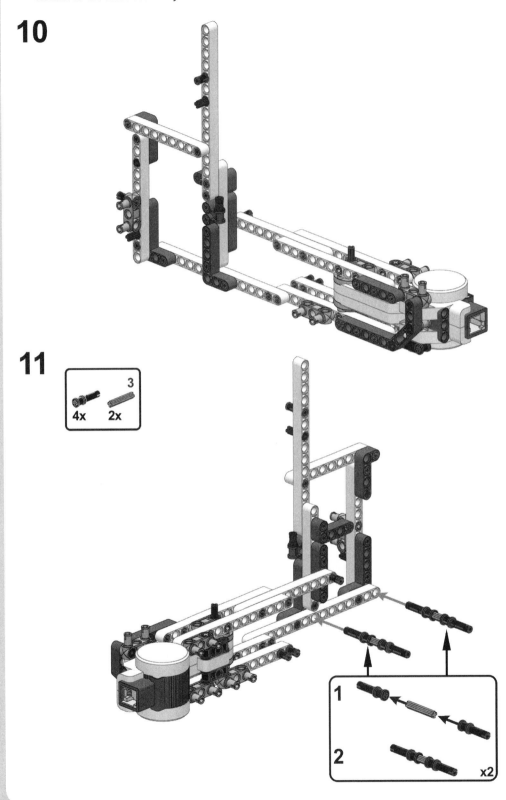

12

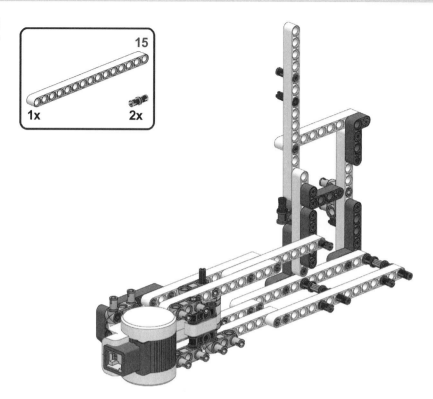

13

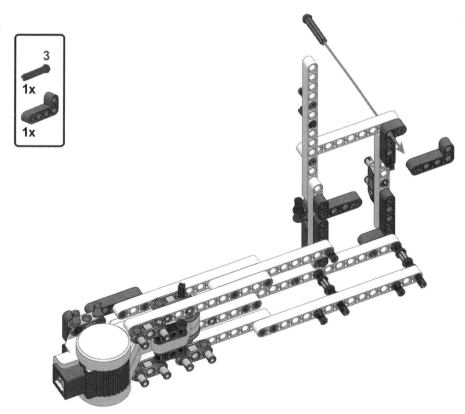

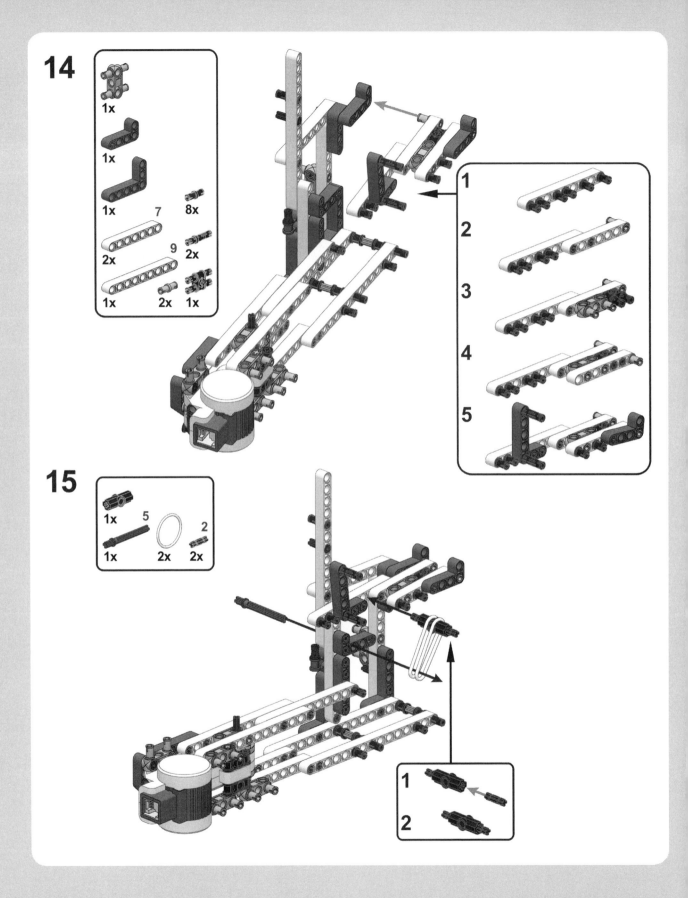

16

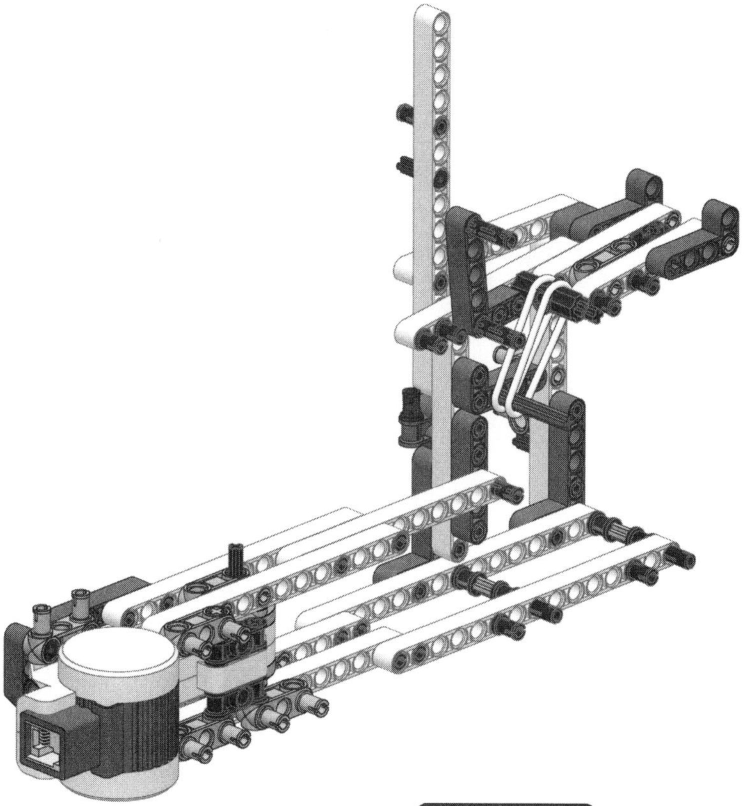

hand

1

5

1x   1x

2

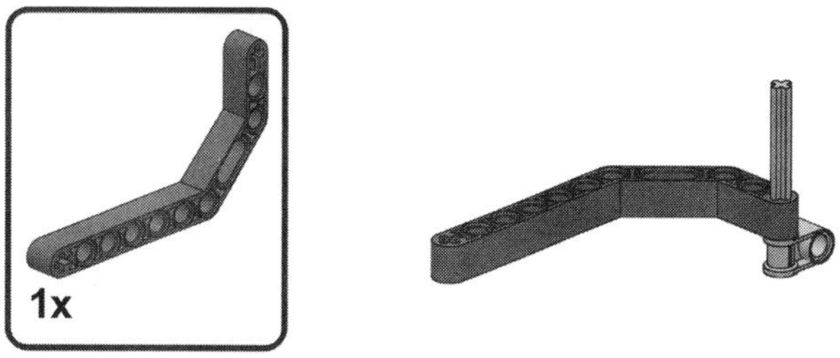

1x

3

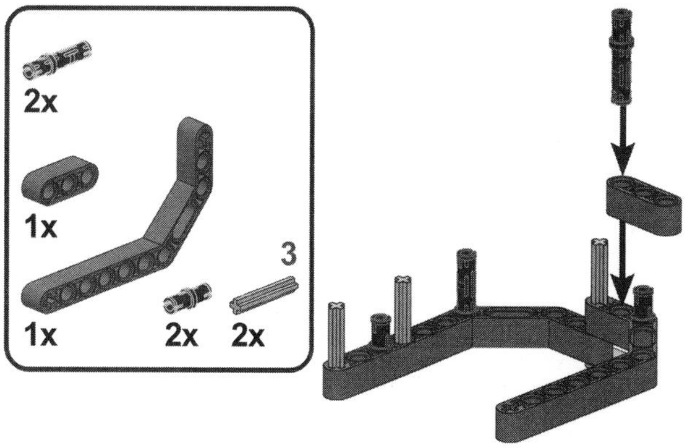

2x

1x

3

1x   2x   2x

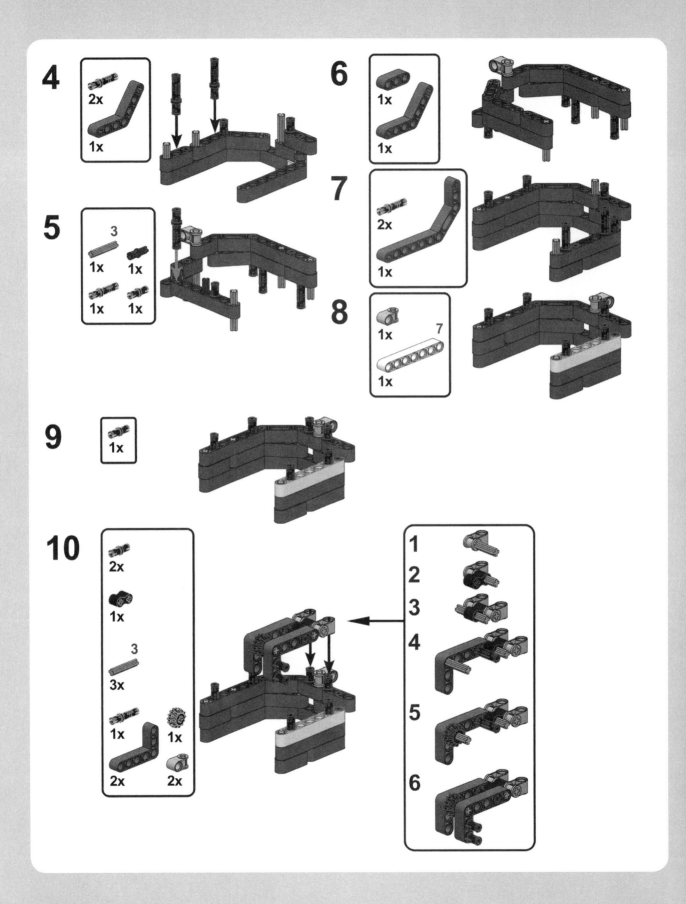

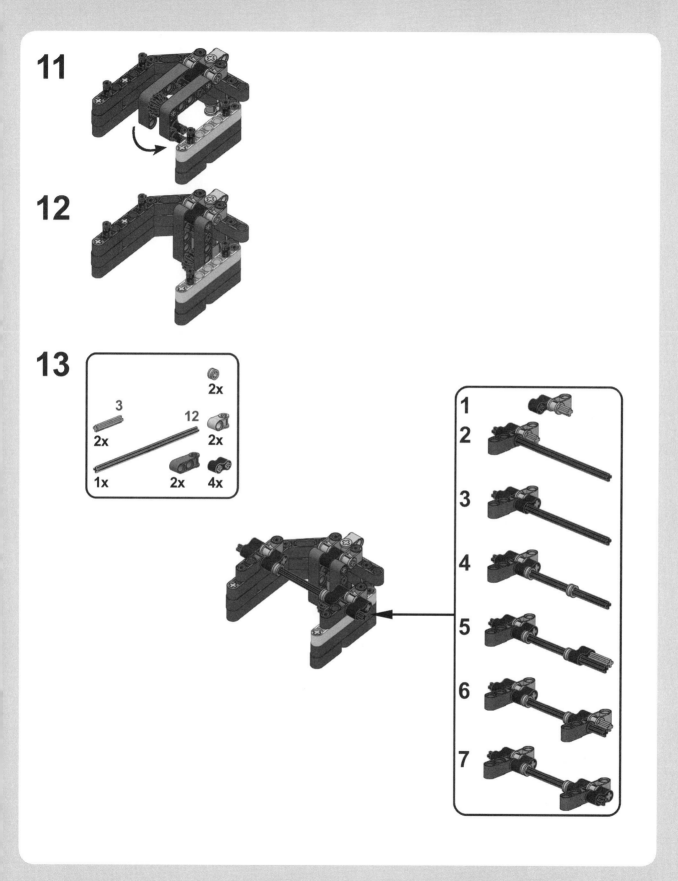

Return to the main assembly.

17

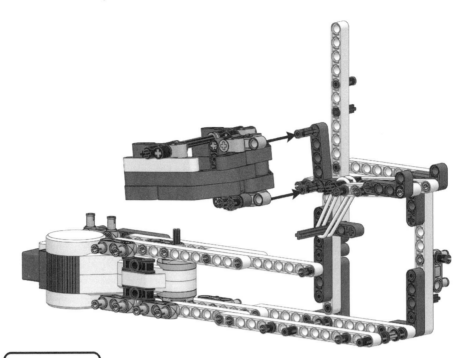

18

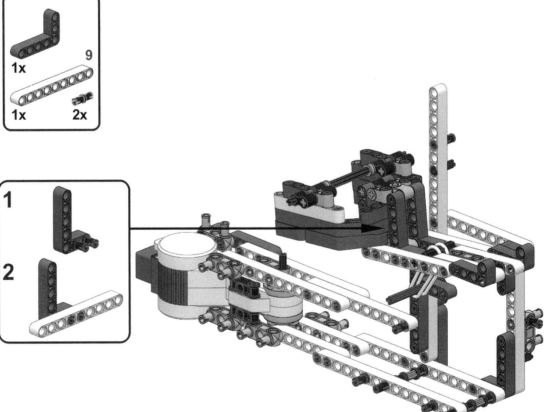

19

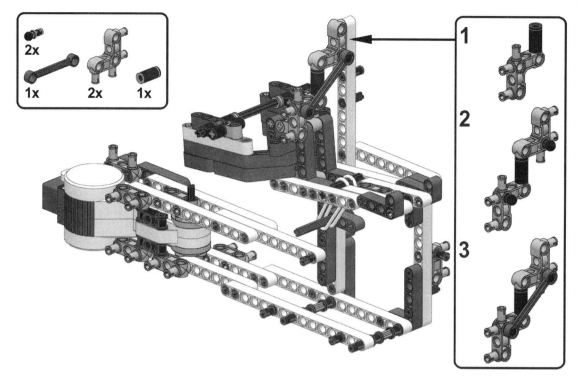

tower 3

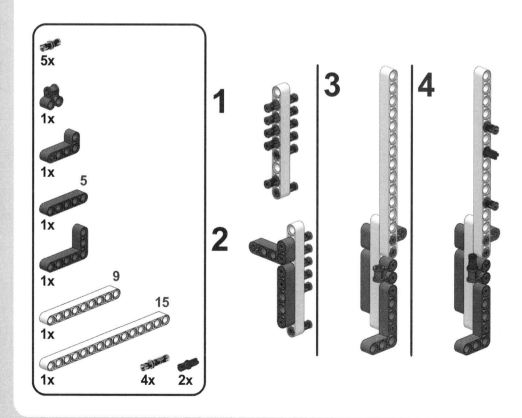

Return to the main assembly.

20

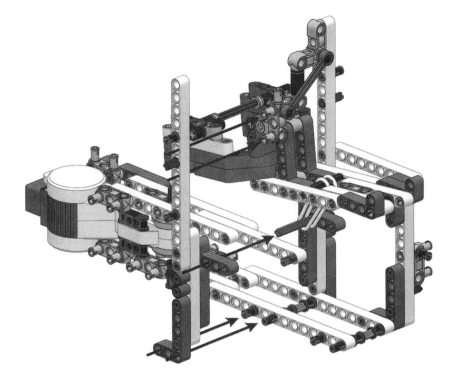

21 1x

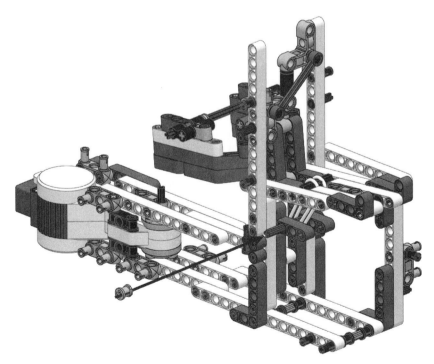

22

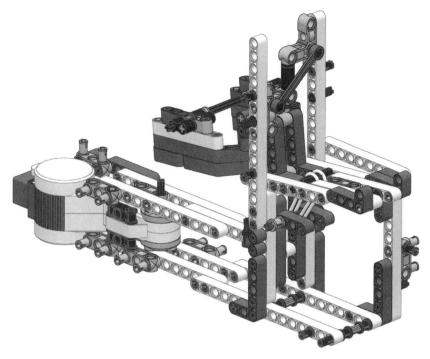

tower 4

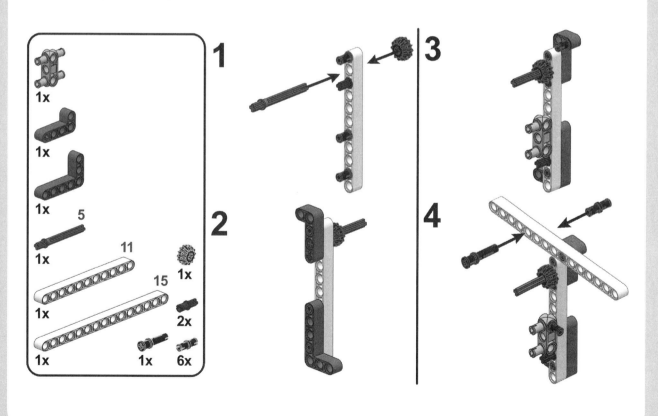

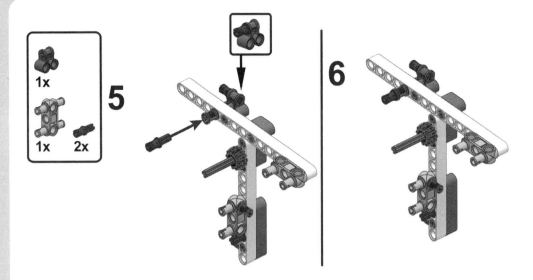

Return to the main assembly.

23

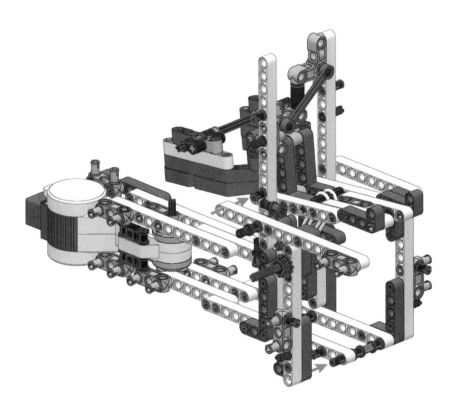

24

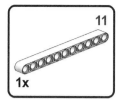

11

1x

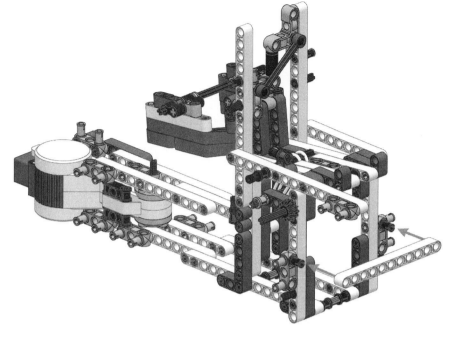

25

15

4x

2x   4x   6x

1

1

2

2

1

1   **2**

2

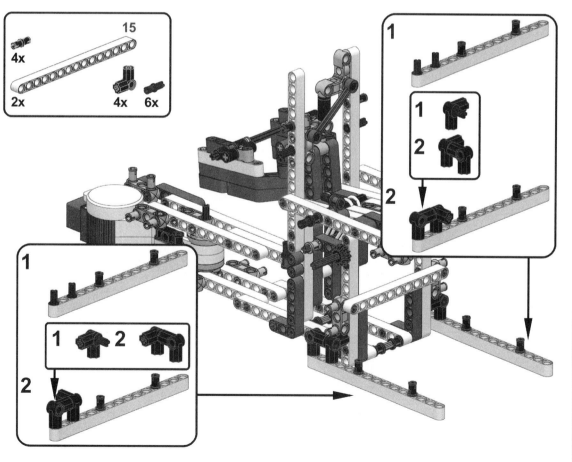

26

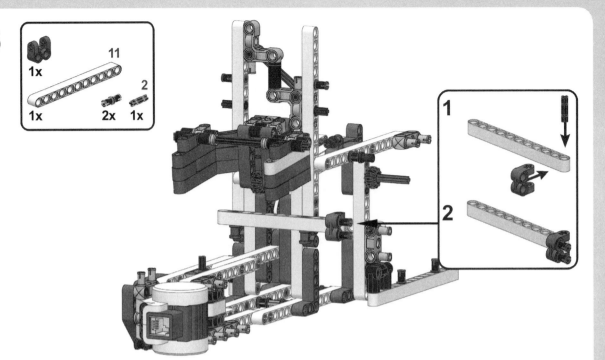

webcam frame 1

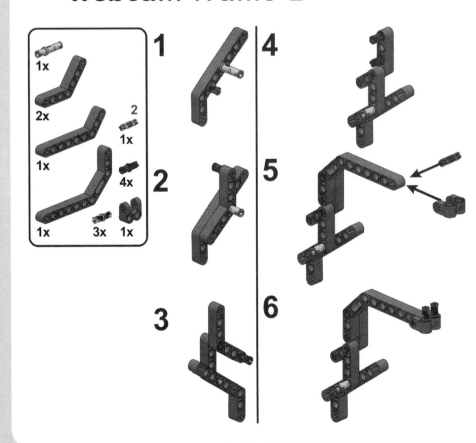

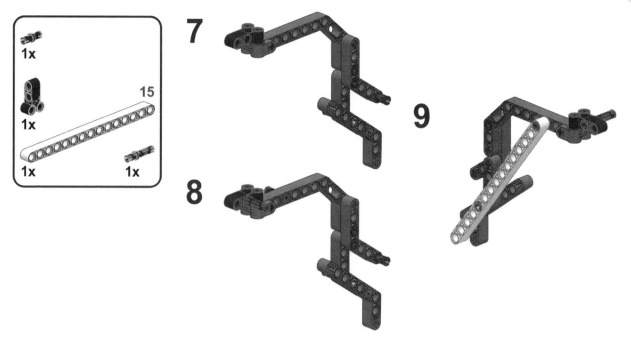

7

8

9

Return to the main assembly.

27

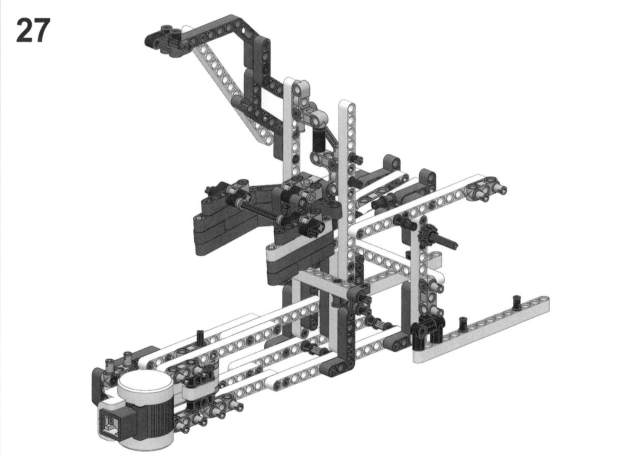

28

2x

2x

2x

5
1x

7
1x

1

2

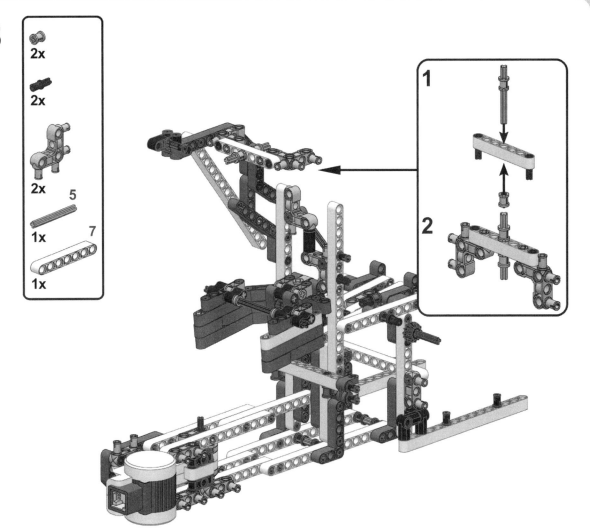

29

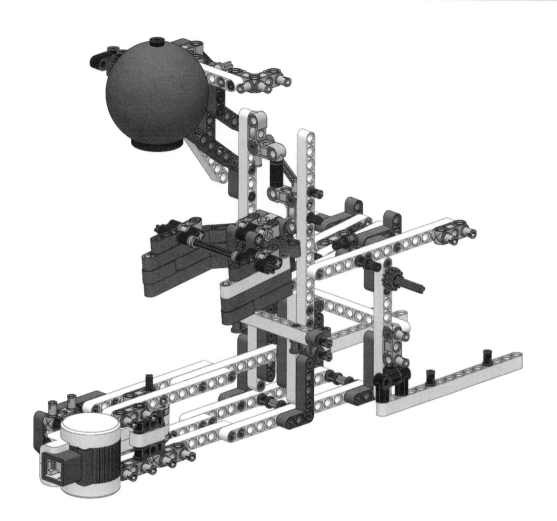

webcam frame 2

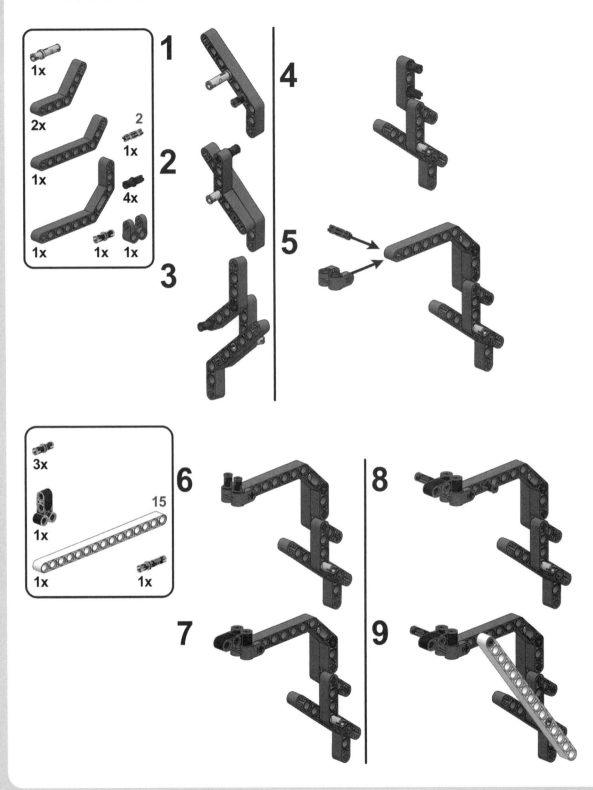

Return to the main assembly.

30

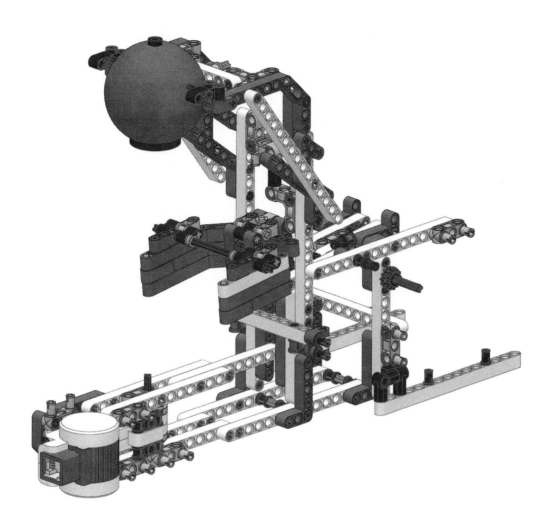

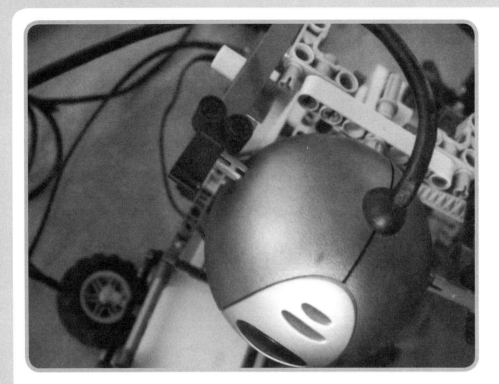

Fit the long pin into the camera's side hole.

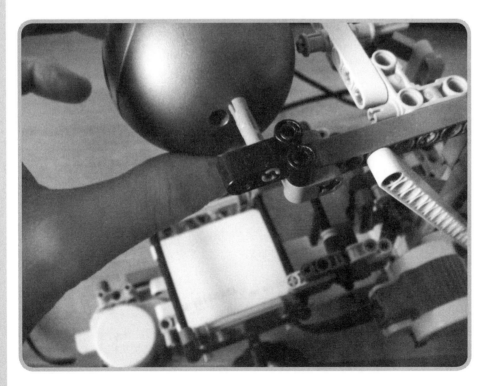

Here's the hole before the pin is inserted.

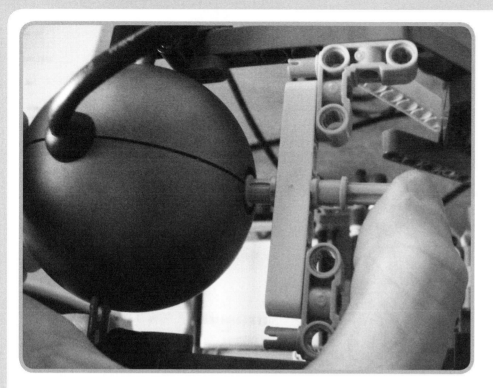

Slide the axle into the bottom hole.

The camera should now be locked in place. Adjust its position slightly while using the CubeSolver program's Video Window to align the facelets with the capture window.

31

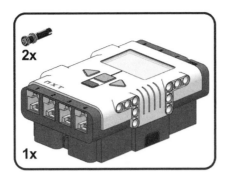

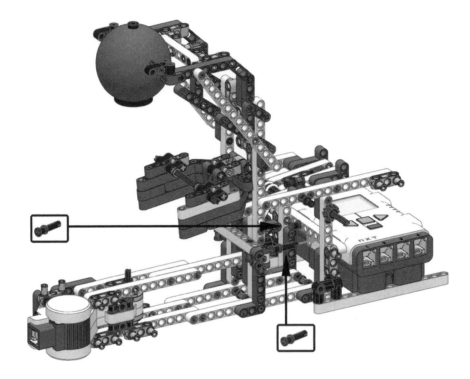

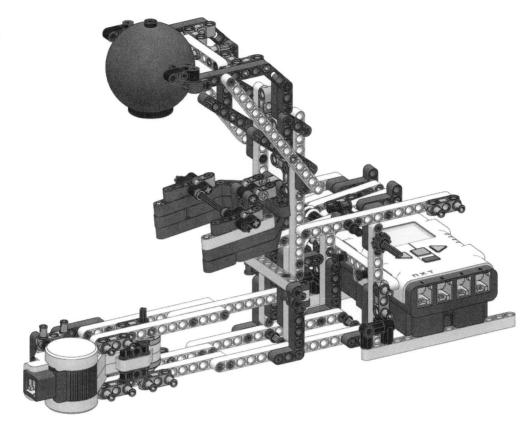

arm motor

1

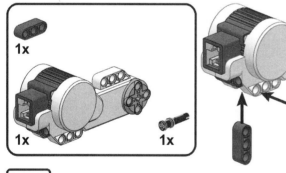

2

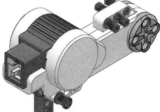

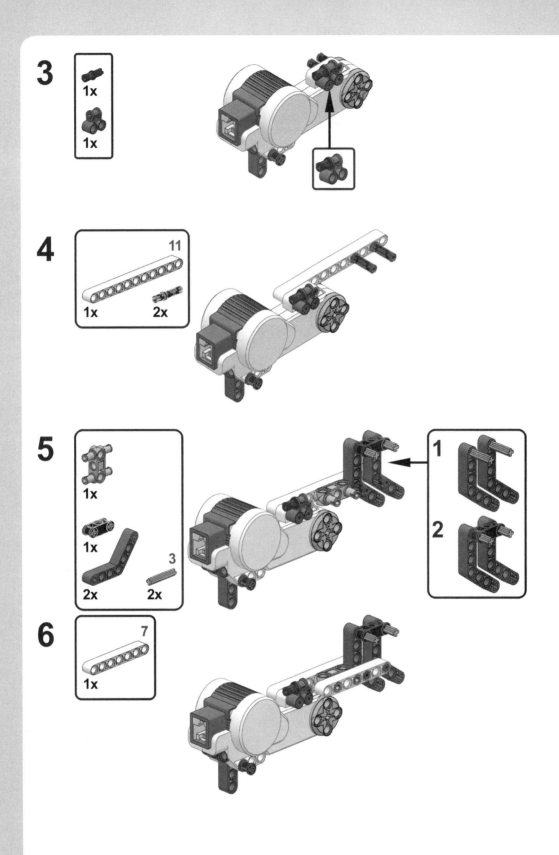

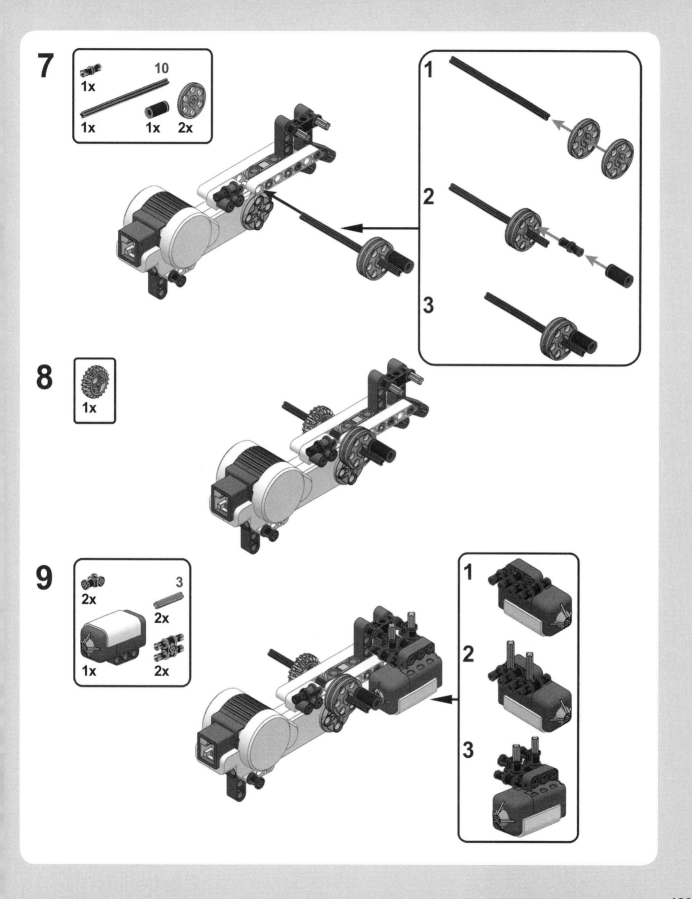

Return to the main assembly.

33

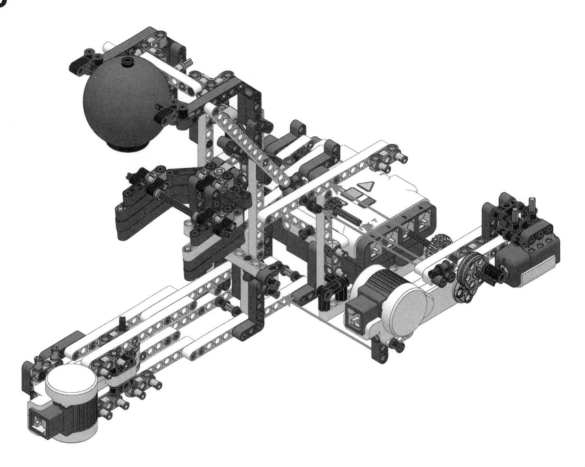

34

1x

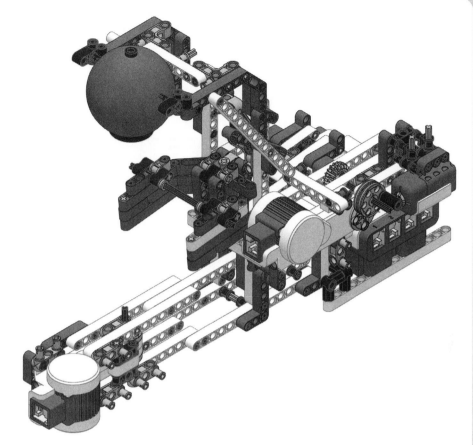

rotating base

1

1x 4x

2

2x

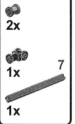

1x

7

1x

3

9

2x 4x

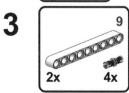

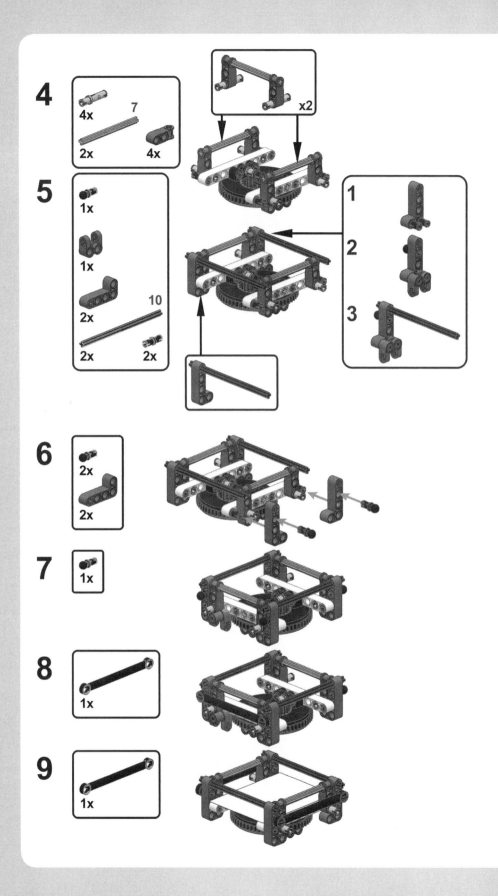

Roll a bit of tape with the adhesive on the outside and attach it to the rotating base as shown here.

Cut a thin piece of cardboard or a thin, rigid, slippery plastic sheet of 65mm × 55mm (2.56" × 2.17").

Attach the sheet to the base as shown here. The cube should be able to slide easily on this surface.

Return to the main assembly.

35

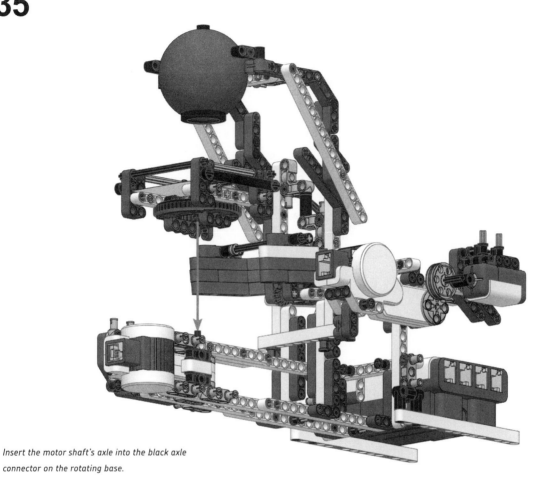

Insert the motor shaft's axle into the black axle connector on the rotating base.

36

37

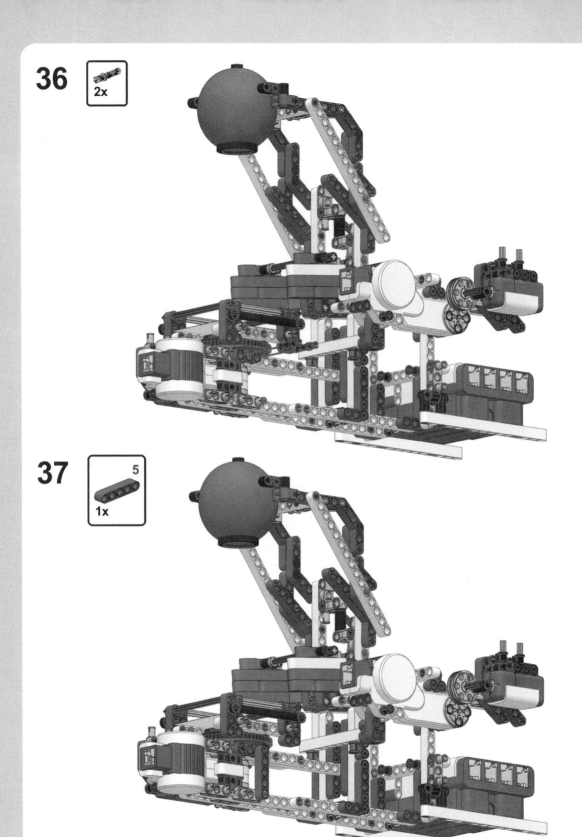

38

39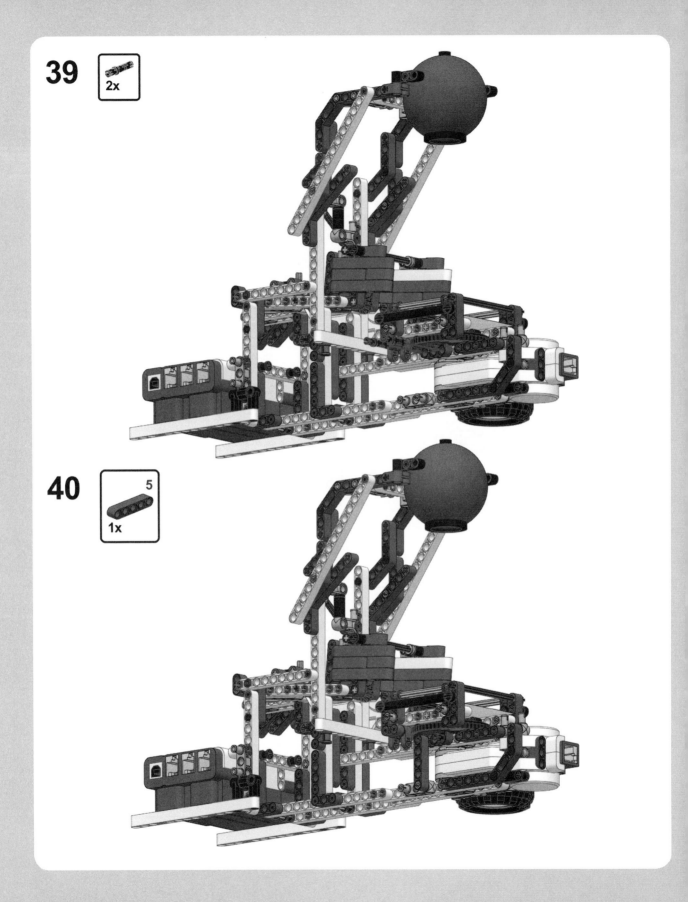

40

41

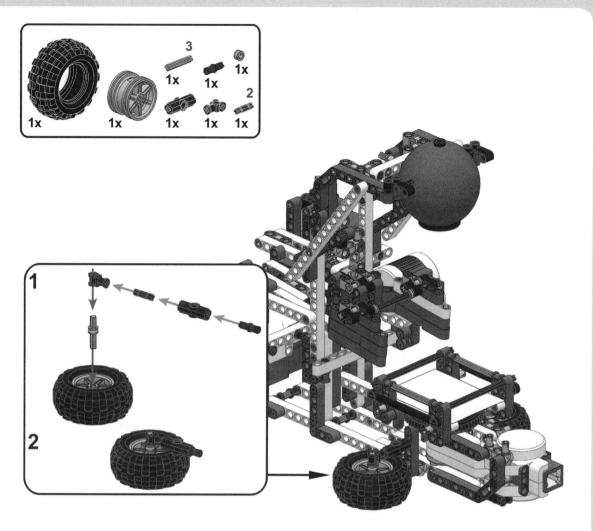

42

20cm

1x

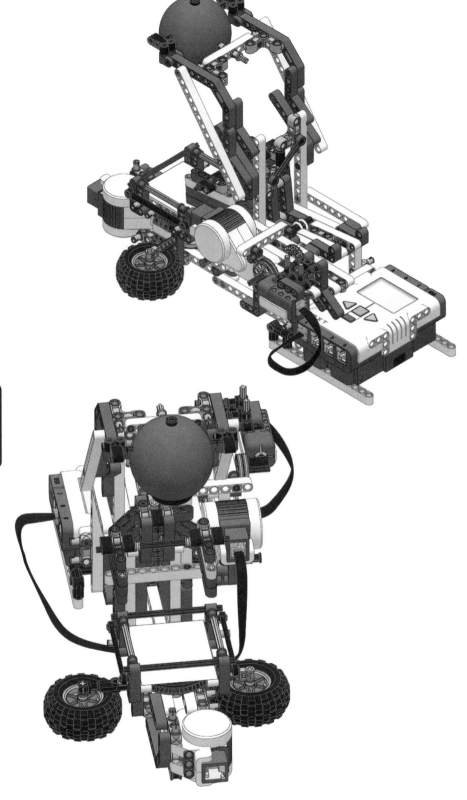

*Connect the touch sensor to IN 1
using a 20cm (8") cable.*

43

35cm

1x

*Connect the arm motor to OUT B
using a 35cm (14") cable.*

44

50cm
1x

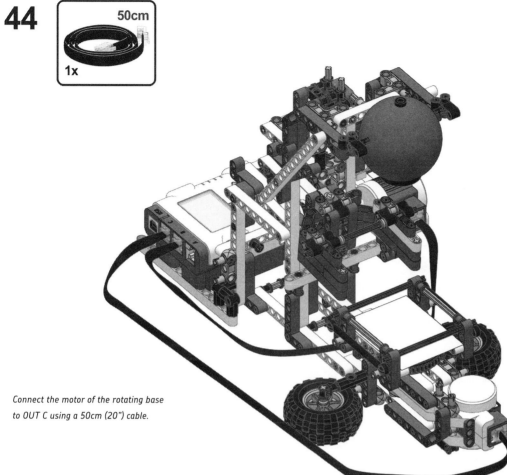

Connect the motor of the rotating base
to OUT C using a 50cm (20") cable.

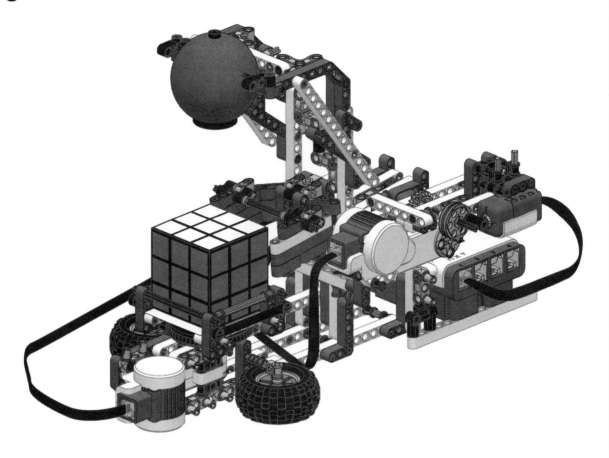

The One-Armed Wonder is ready to solve the Rubik's Cube!

7

building the One-Armed Wonder with the NXT 2.0 set

In this chapter you will build the One-Armed Wonder using the 8547 NXT 2.0 set. In addition to the parts in your set, you will need a USB webcam (as discussed in Chapters 1 and 5) and a 65mm × 55mm (2.56" × 2.17") sheet of slippery cardboard or plastic to put on the rotating base. See pages 134–135 for information on attaching the webcam, and pages 143–144 for information on attaching the slippery sheet.

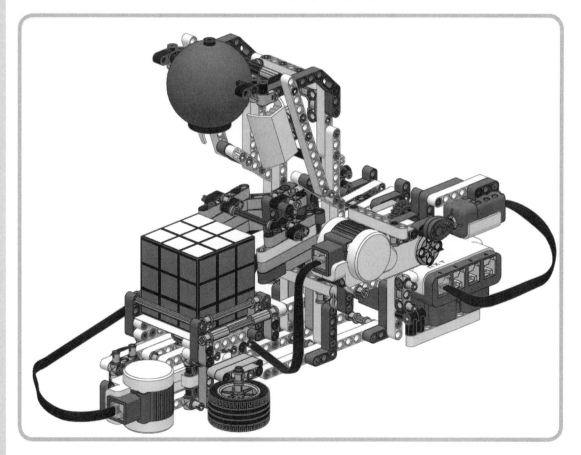

bill of materials

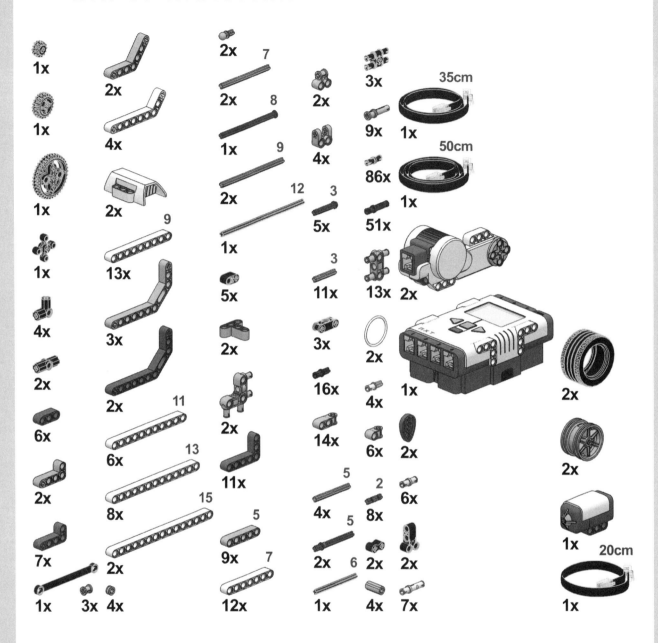

main assembly

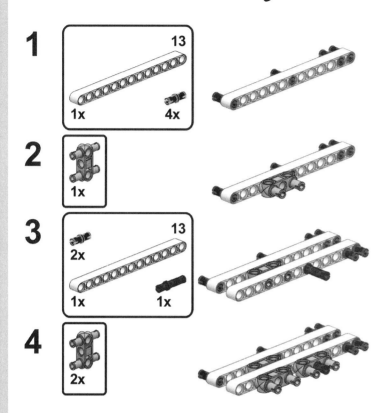

1 13
1x   4x

2
1x

3 13
2x
1x   1x

4
2x

base motor

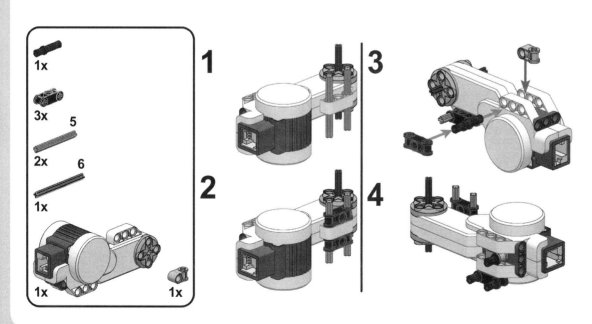

1x

3x   5

2x   6

1x

1x   1x

1

2

3

4

Return to the main assembly.

5 1x

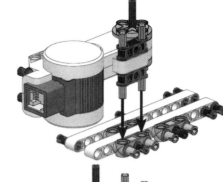

6 1x

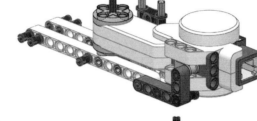

7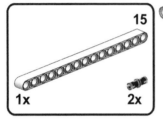

base top layer

1

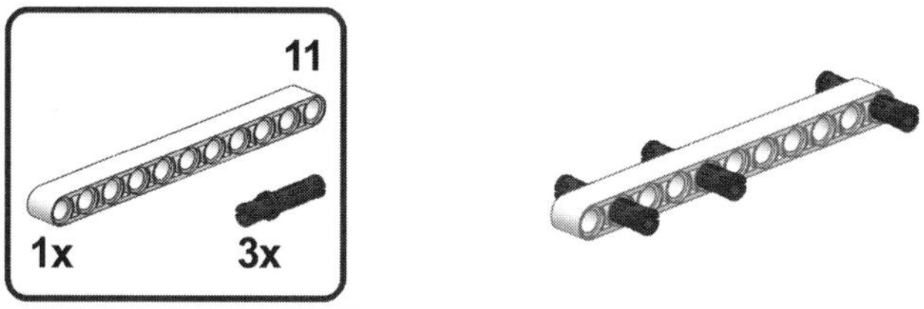

11
1x   3x

2

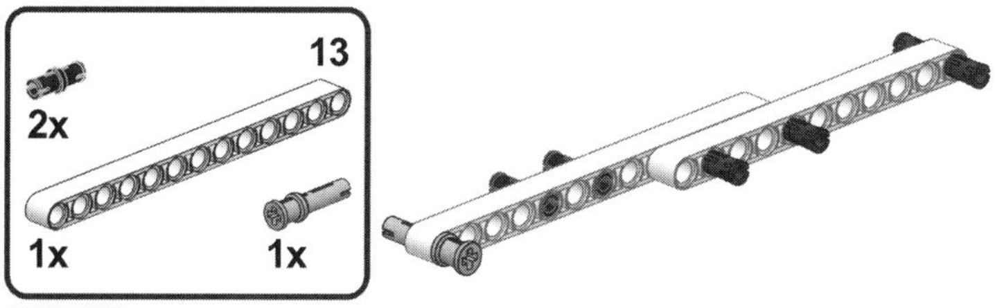

2x
13
1x   1x

3

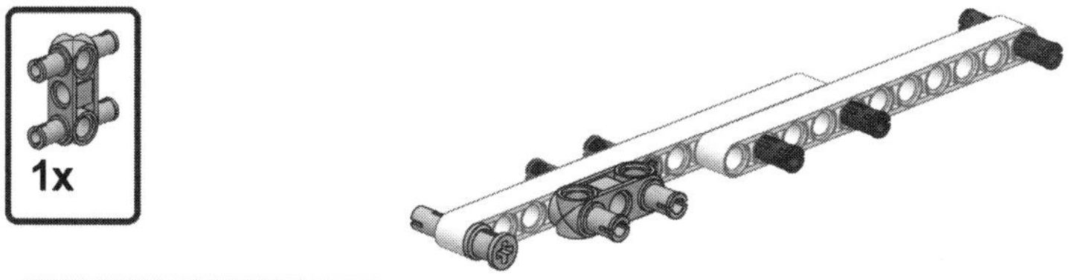

1x

4

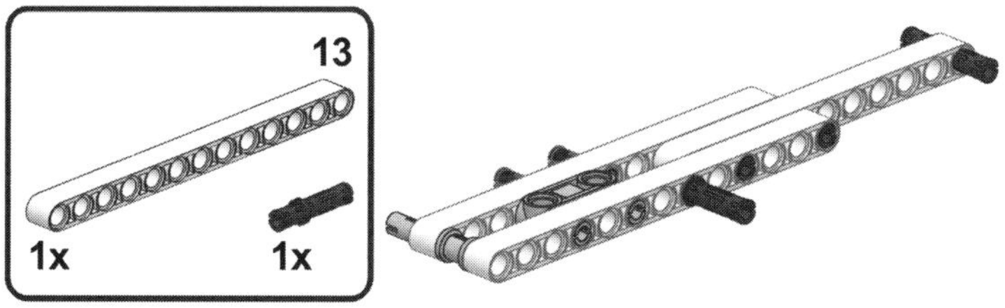

13
1x   1x

5

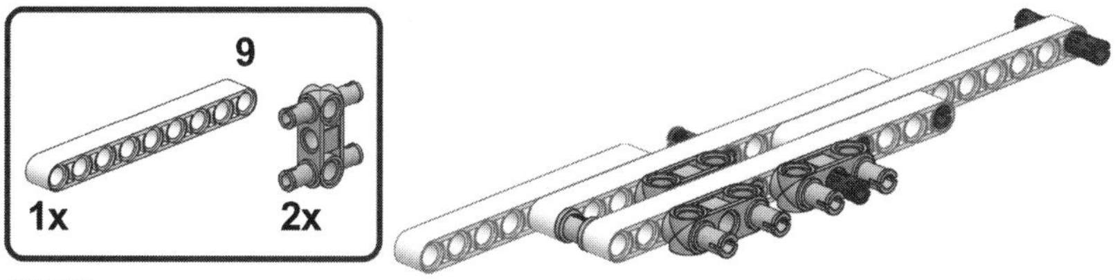

9
1x   2x

6

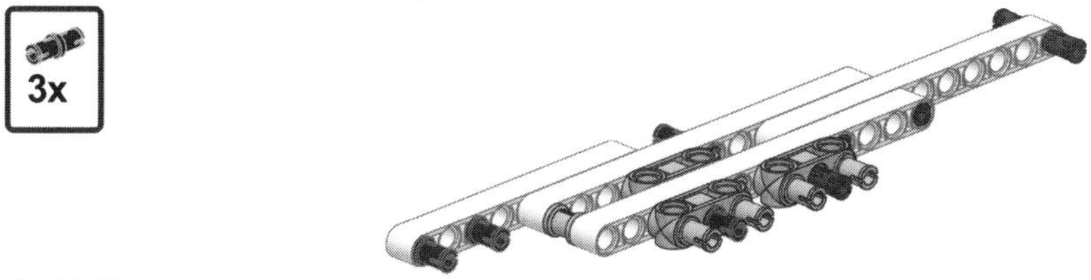

3x

7

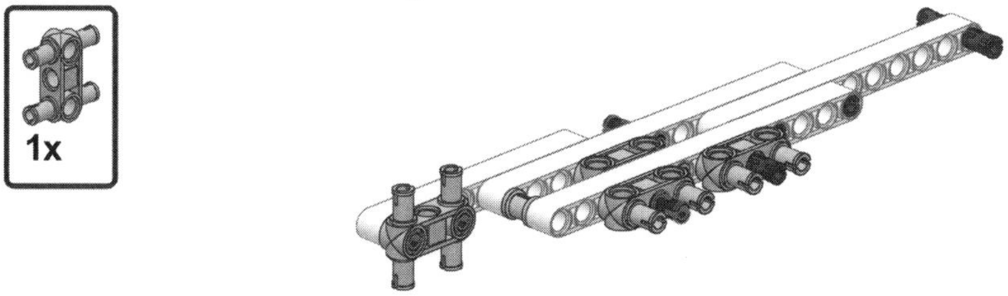

1x

Return to the main assembly.

8

9

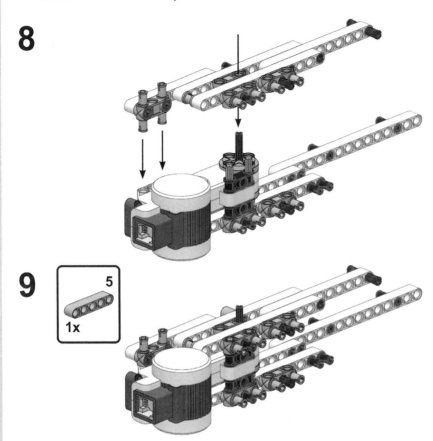

tower 1

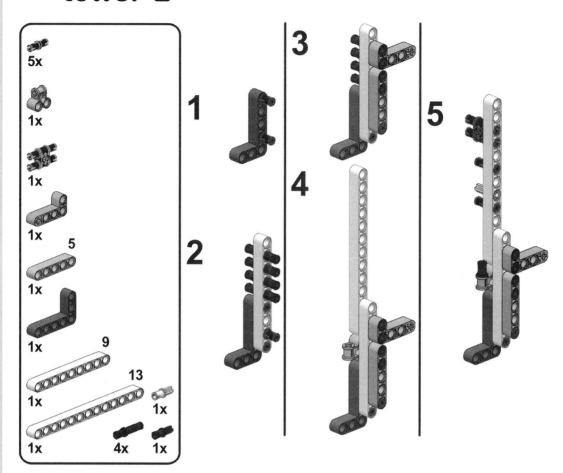

Return to the main assembly.

10

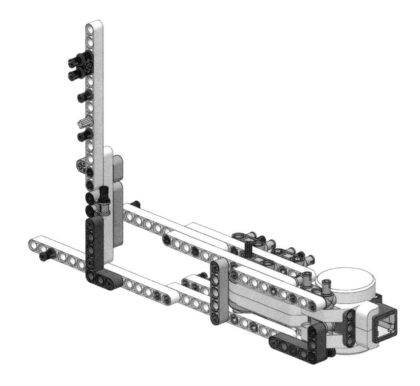

tower 2

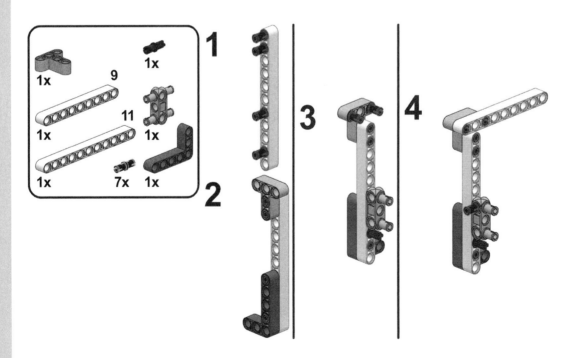

Return to the main assembly.

11

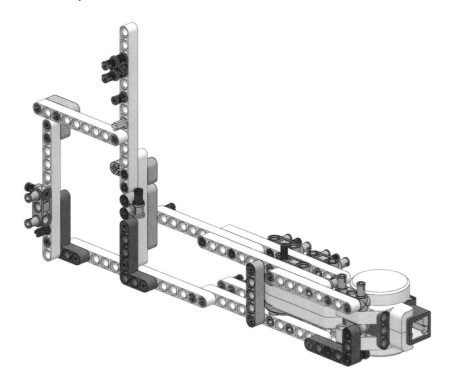

12

5
2x

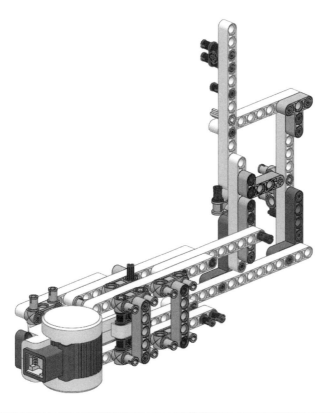

13

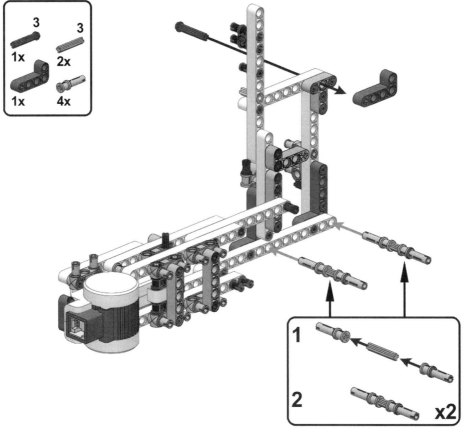

1
2

x2

14

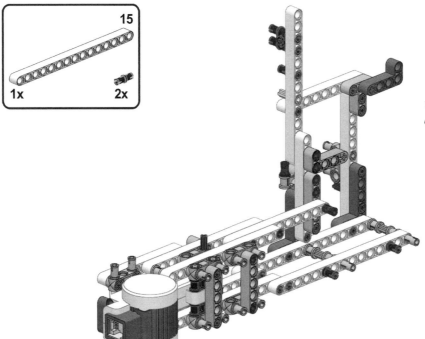

The 2 × 4 liftarm is hanging down freely in reality.

arm

1
9
1x   6x

2
7
1x   1x

3
1x   1x

4
7
1x   1x

5
2x
1x   1x   2x

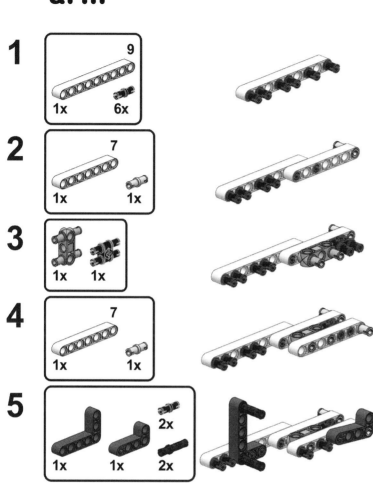

Return to the main assembly.

15

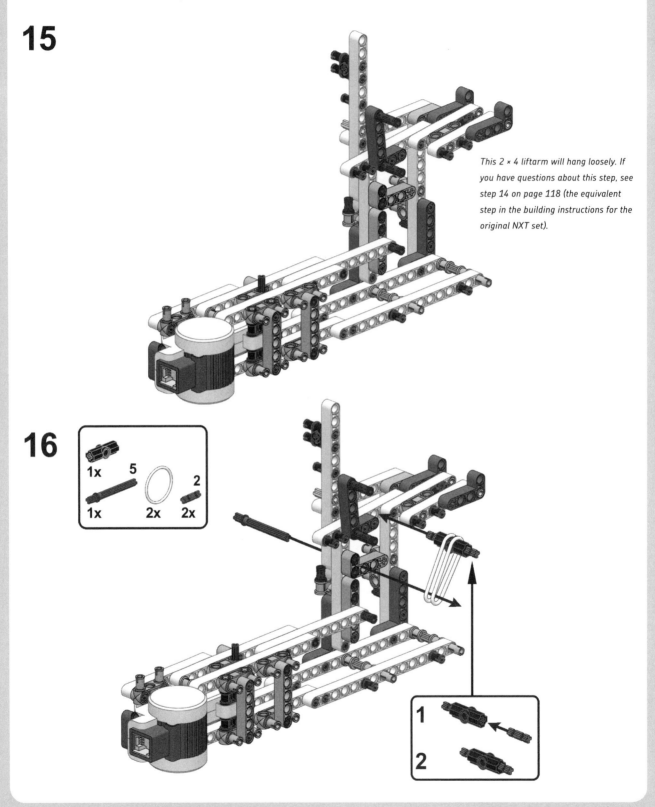

This 2 × 4 liftarm will hang loosely. If you have questions about this step, see step 14 on page 118 (the equivalent step in the building instructions for the original NXT set).

16

1x   5

1x   2x   2x 2

1

2

17

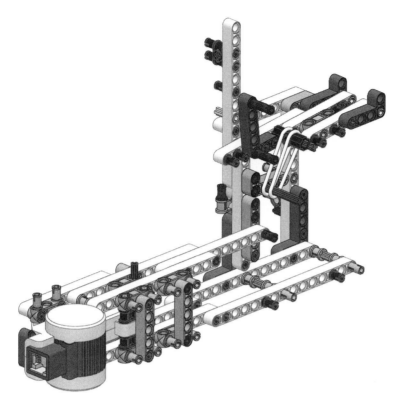

hand

1

5

1x 1x

2

1x

3

2x

1x

1x 2x 2x 3

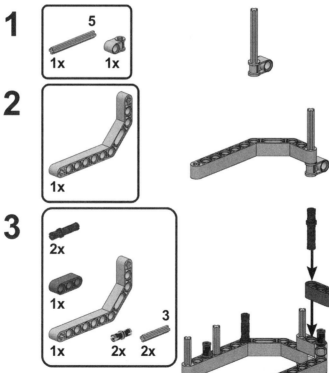

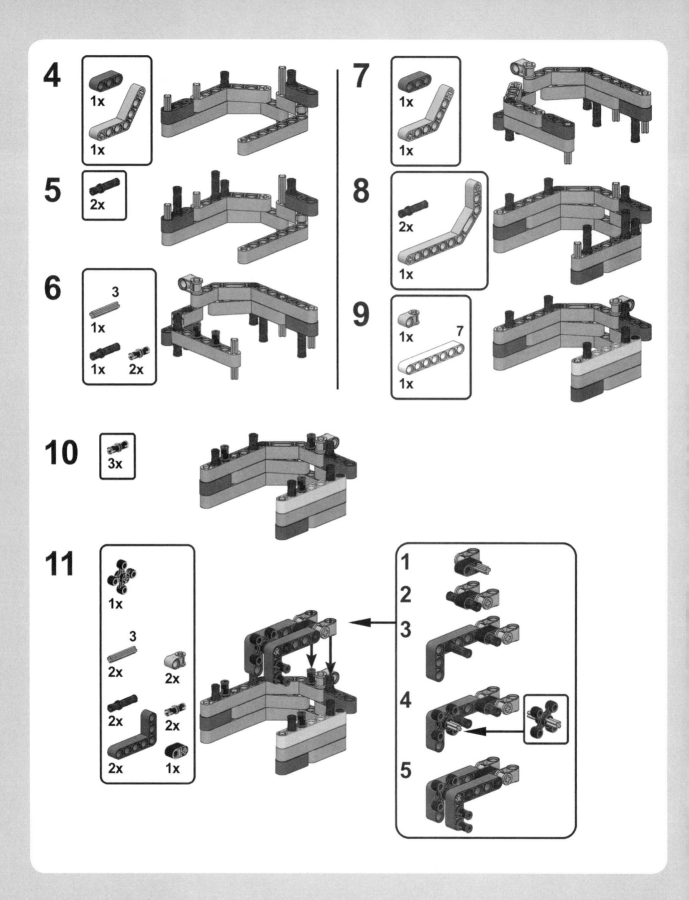

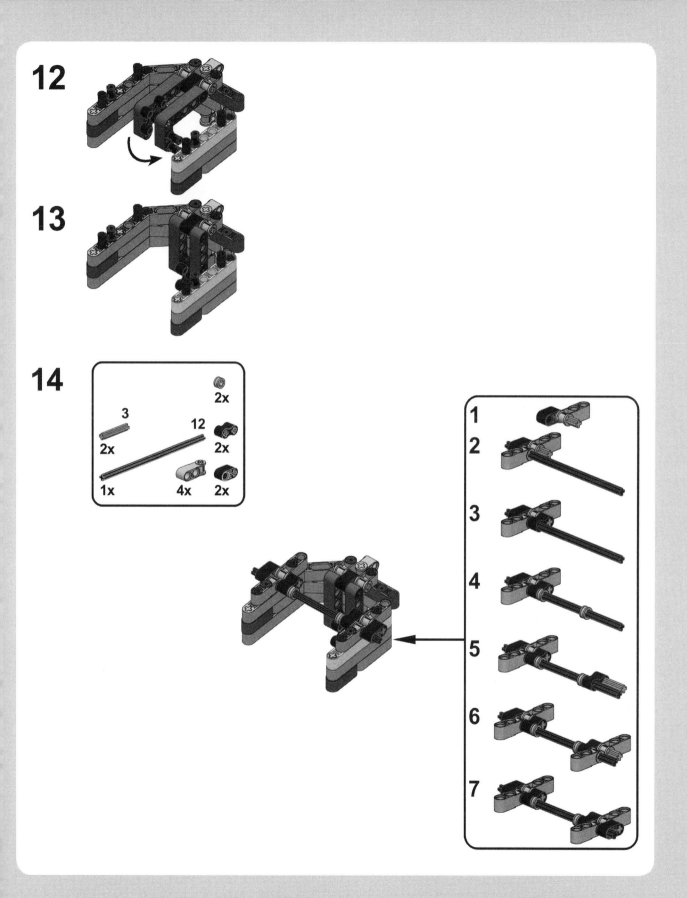

Return to the main assembly.

18

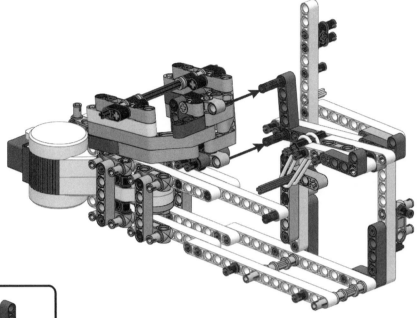

19

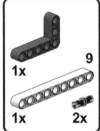

1x 9
1x 2x

1
2

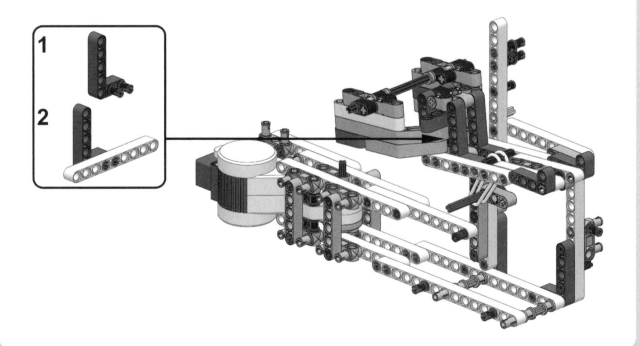

tower 3

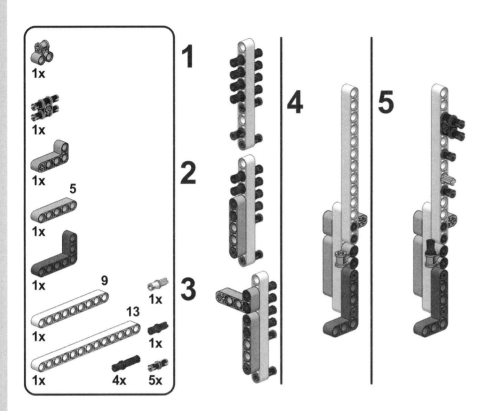

1
2
3

4
5

Return to the main assembly.

20

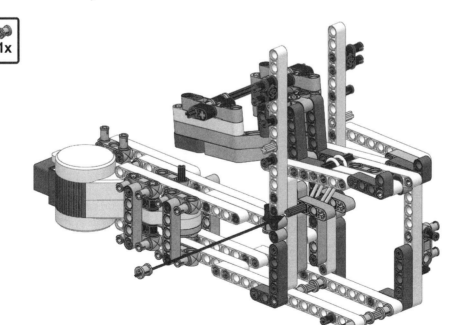

21

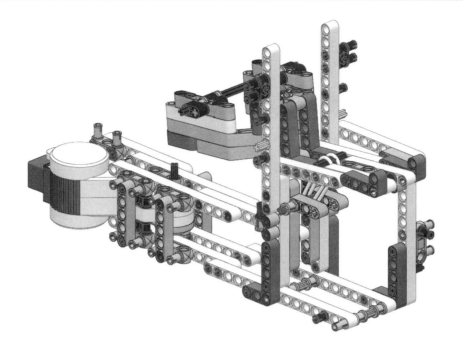

tower 4

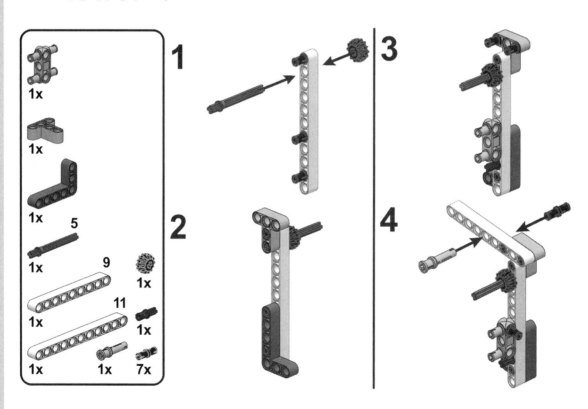

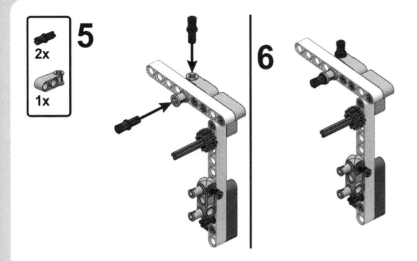

Return to the main assembly.

22

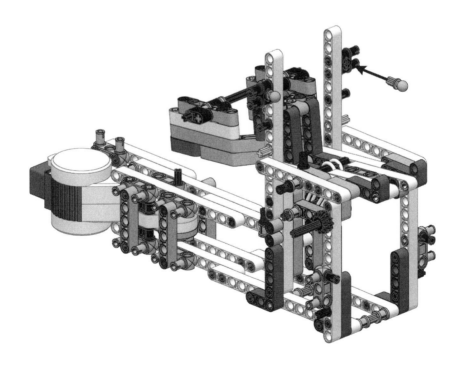

23

11
1x
1x

24

9
4x
4x
2x
2x
4x

1

1
2

2

1

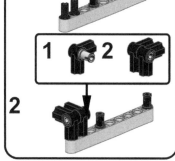

1
2

2

25

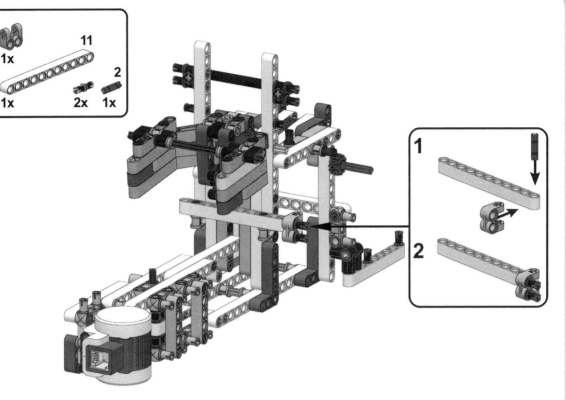

webcam frame 1

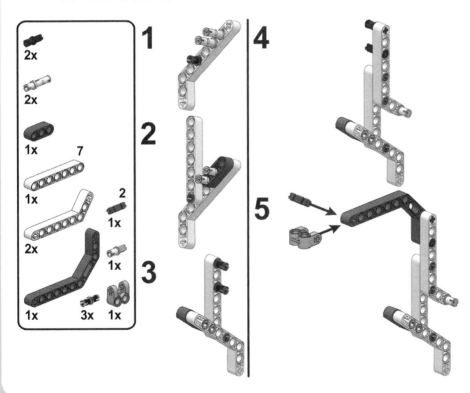

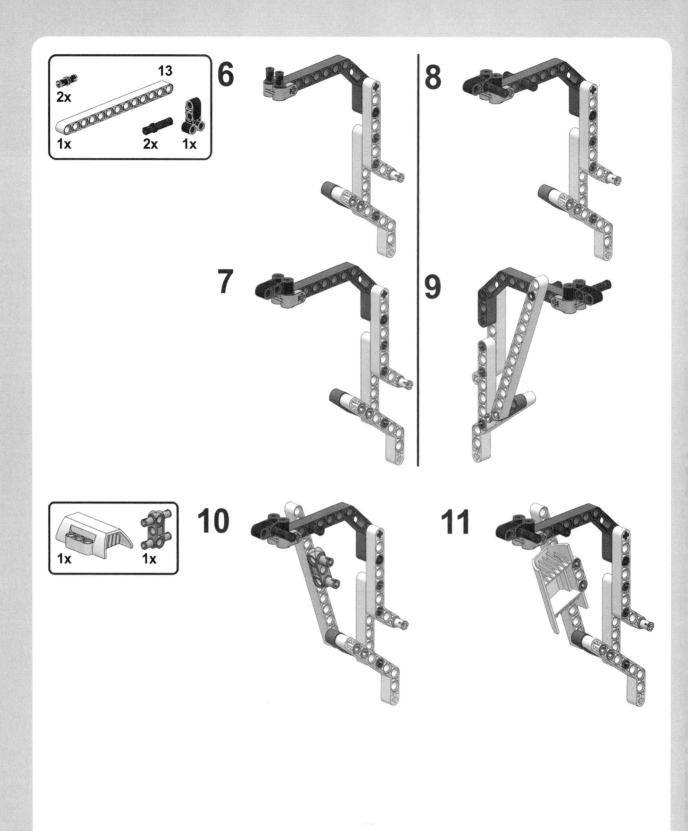

Return to the main assembly.

26

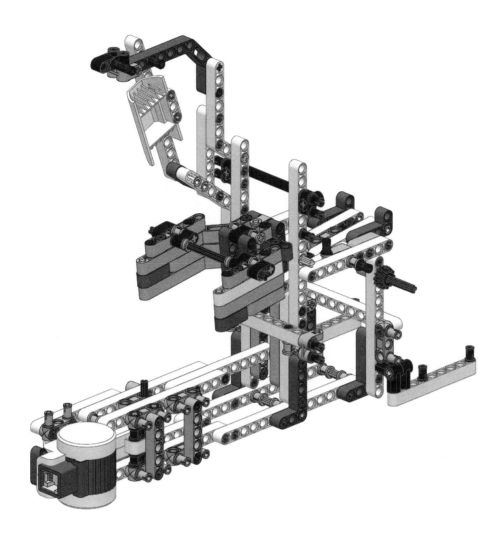

27

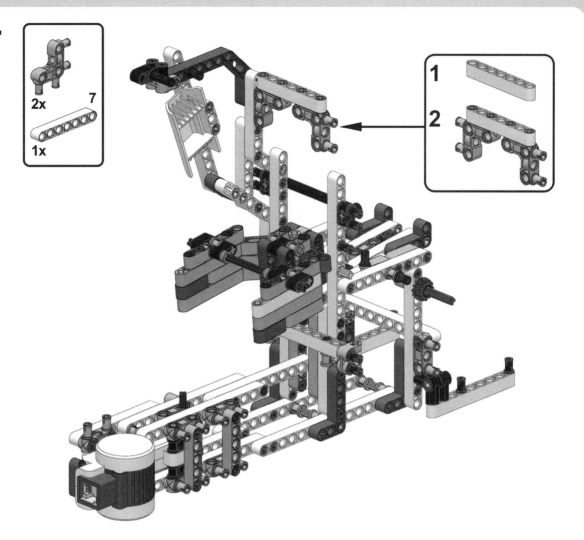

28

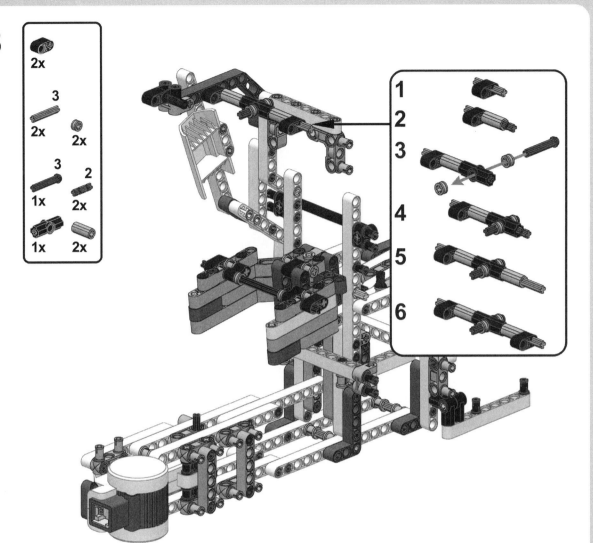

29

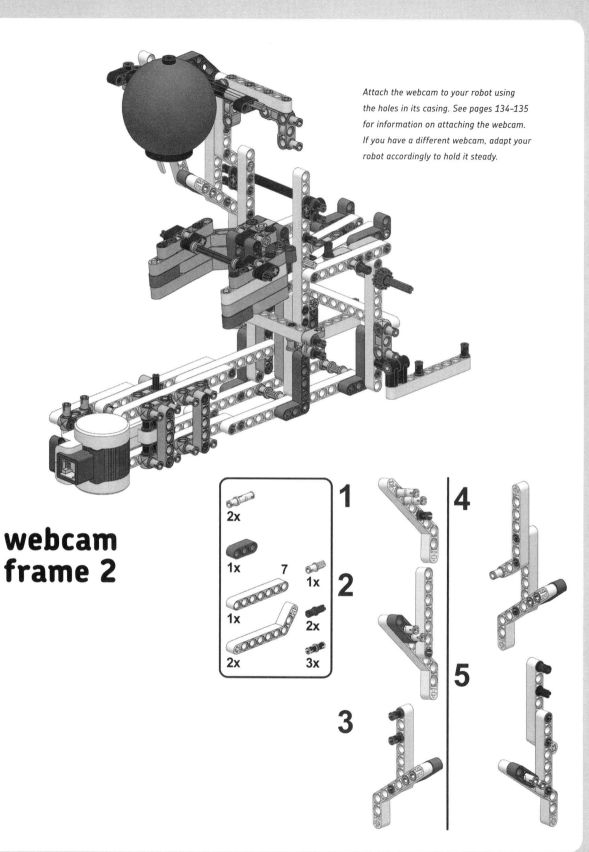

Attach the webcam to your robot using the holes in its casing. See pages 134–135 for information on attaching the webcam. If you have a different webcam, adapt your robot accordingly to hold it steady.

webcam frame 2

2x

1x

7

1x

1x

2x

2x

3x

1

2

3

4

5

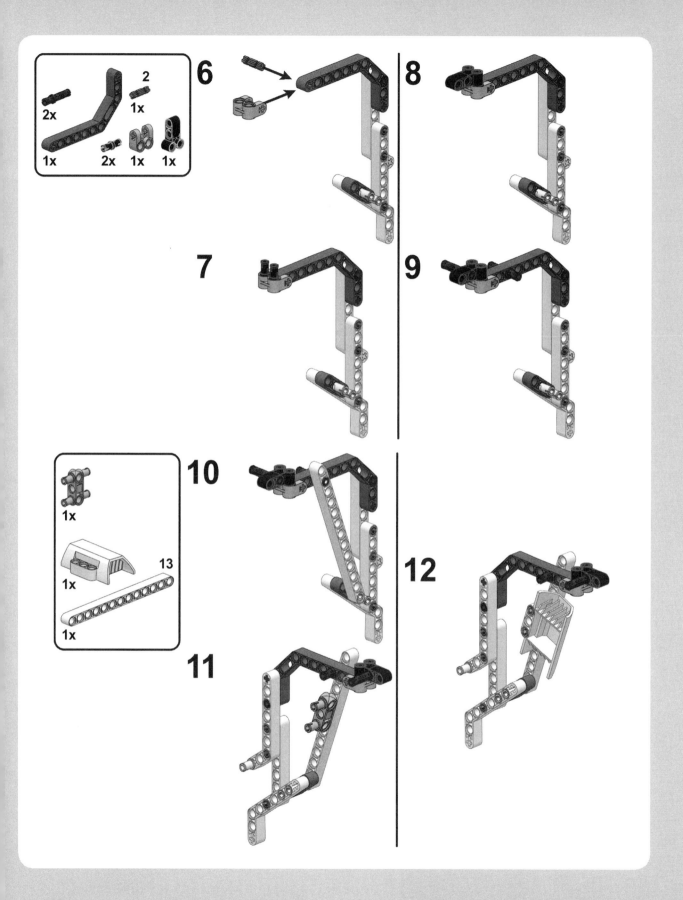

Return to the main assembly.

30

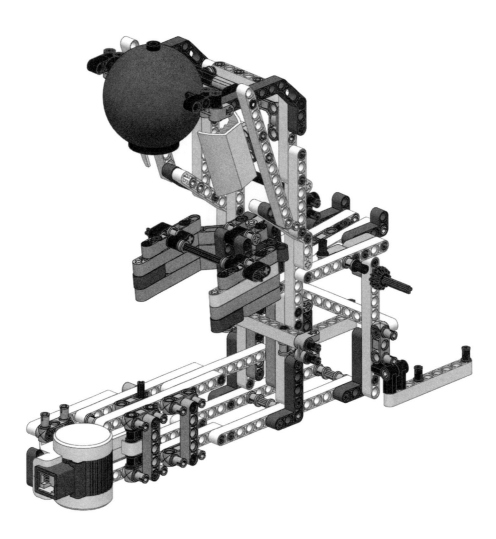

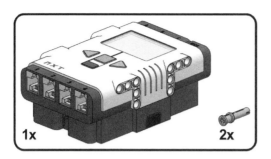

31

1x   2x

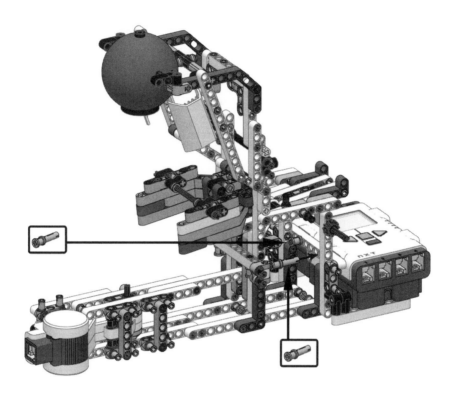

32

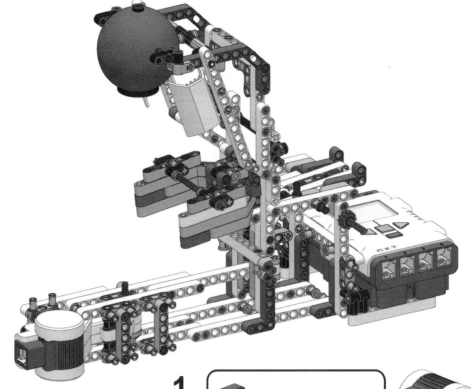

arm motor

1

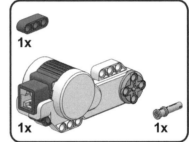

1x

1x   1x

2

1x

2x

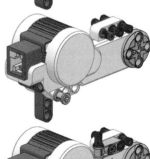

3

1x

1x

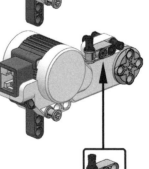

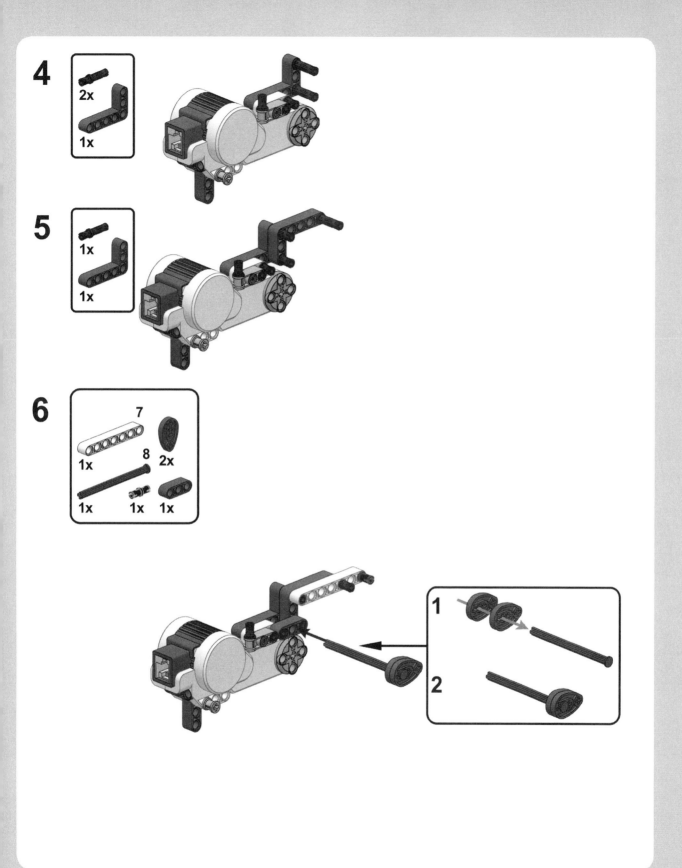

7

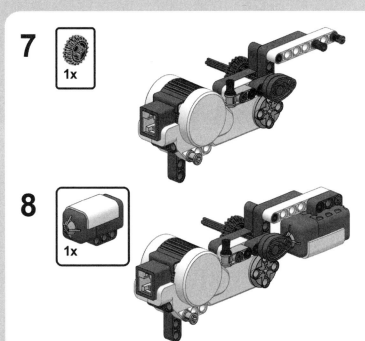

8

Return to the main assembly.

33

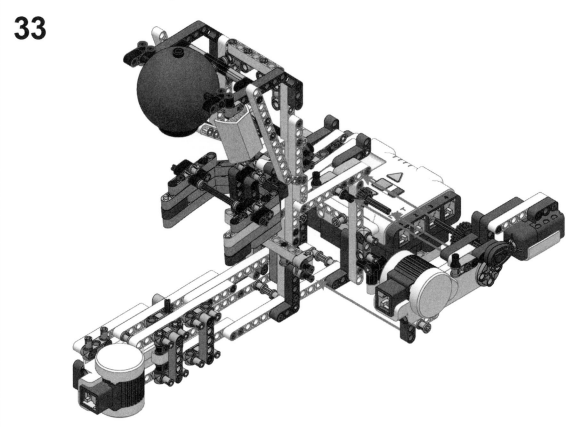

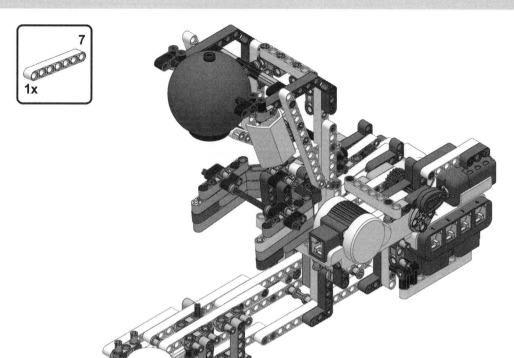

support 1

1

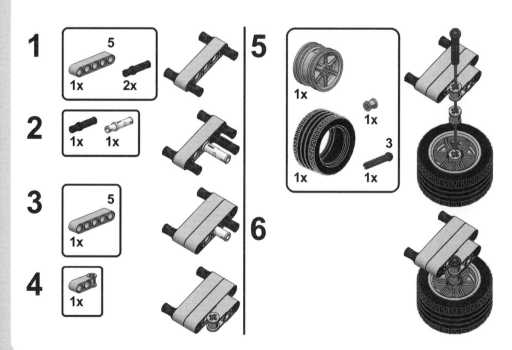

2

3

4

5

6

Return to the main assembly.

35

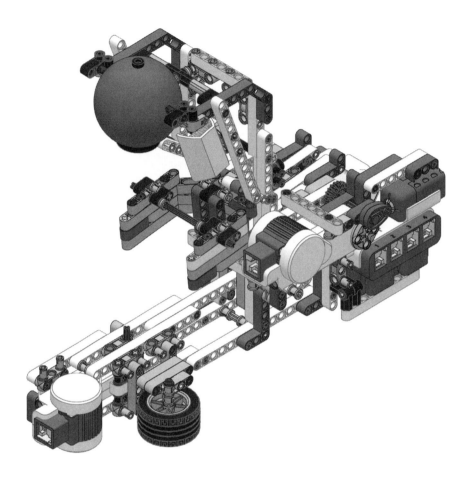

support 2

1 5 1x 3x

2 1x 1x

3 5 1x

4 1x

5 1x 1x 1x 1x 3

6 1x

7 1x

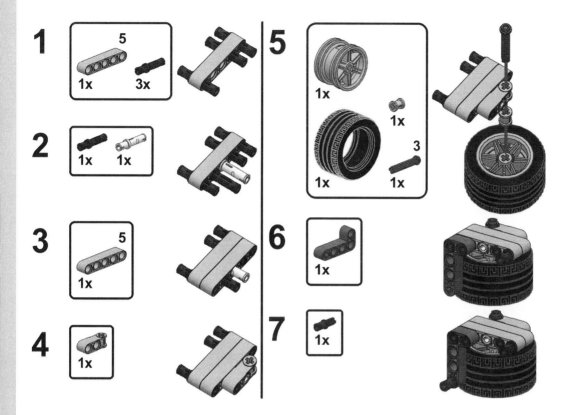

Return to the main assembly.

36

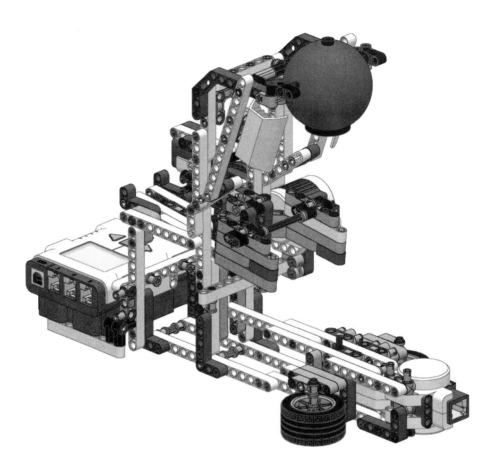

rotating base

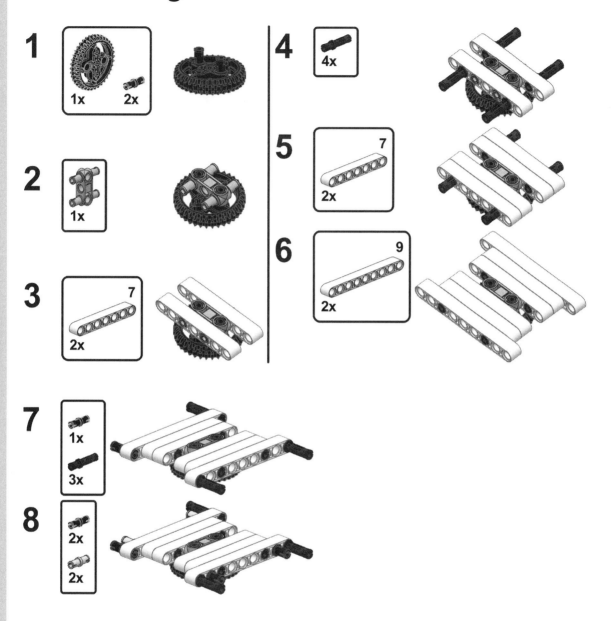

1 1x 2x

2 1x

3 7 2x

4 4x

5 7 2x

6 9 2x

7 1x 3x

8 2x 2x

9

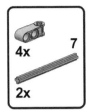

4x

7

2x

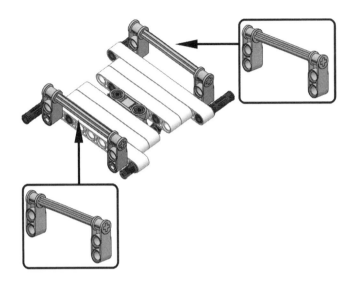

10

1x

9

1x

1x   2x   2x   1x

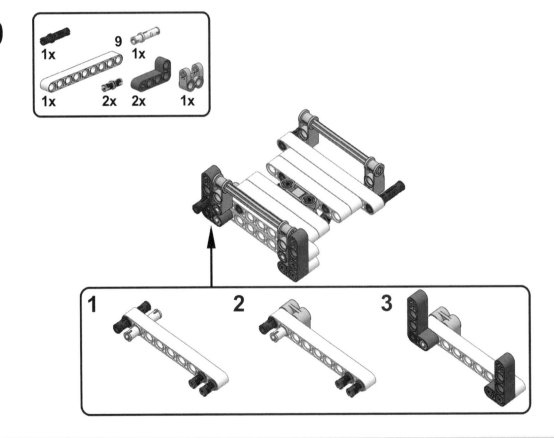

1   2   3

11

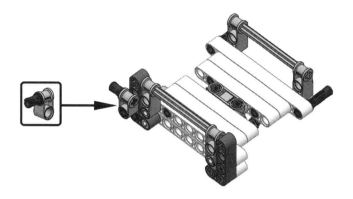

12

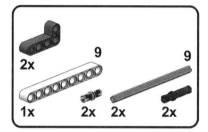

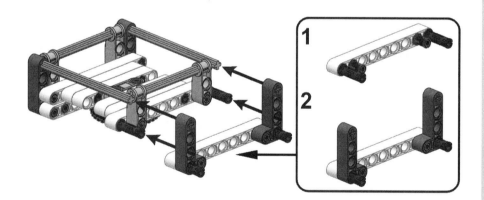

13

1x 5 2 1x

1x 2x

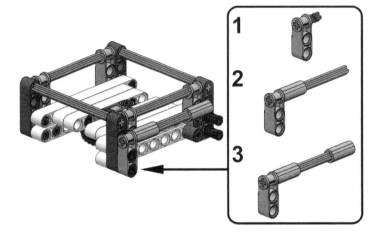

1

2

3

14

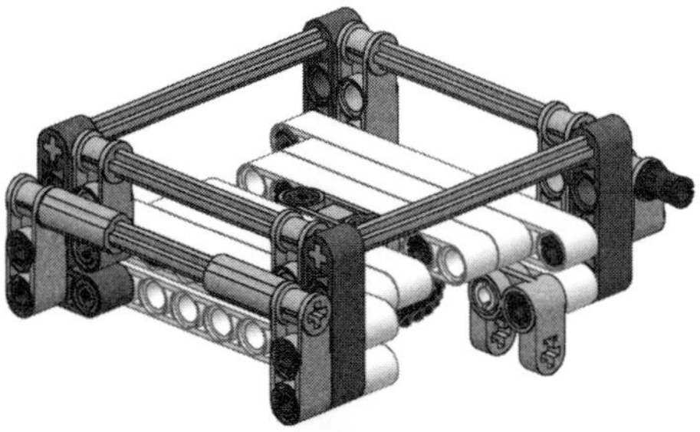

1x

15

3

11

1x

1x

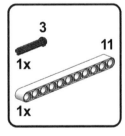

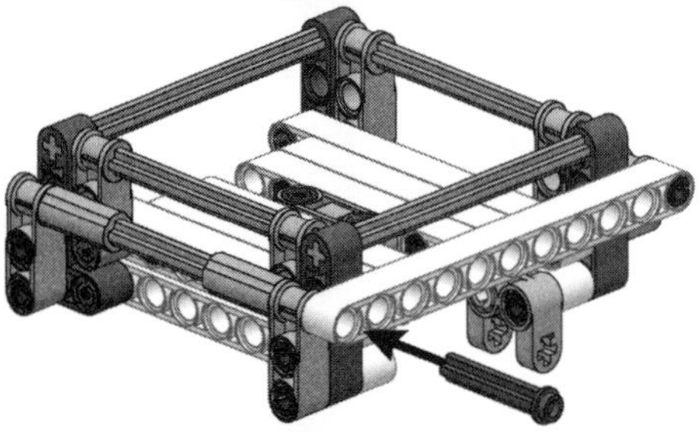

16

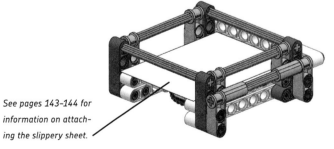

See pages 143–144 for information on attaching the slippery sheet.

Return to the main assembly.

37

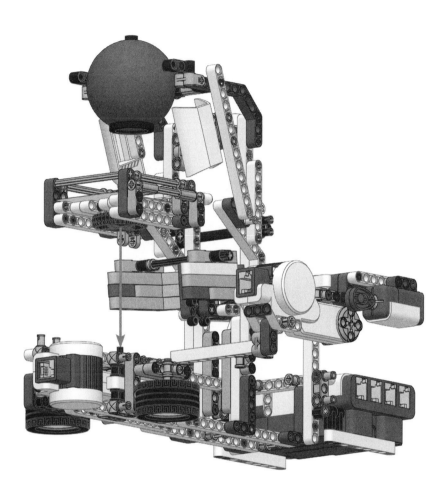

38

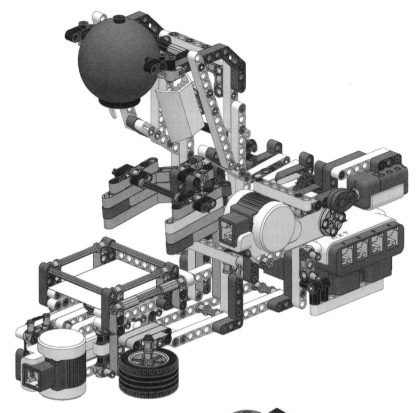

39

20cm

1x

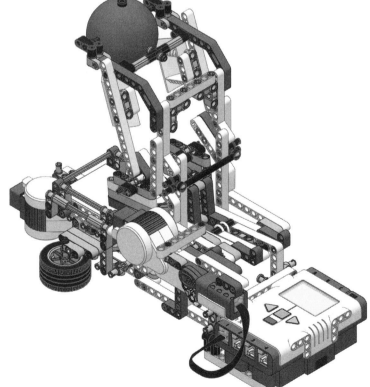

*Connect the touch sensor to IN 1
using a 20cm (8") cable.*

40

35cm

1x

Connect the arm motor to OUT B
using a 35cm (14") cable.

41

50cm

1x

Connect the motor of the rotating base
to OUT C using a 50cm (20") cable.

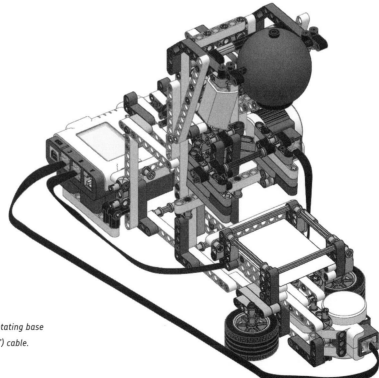

42

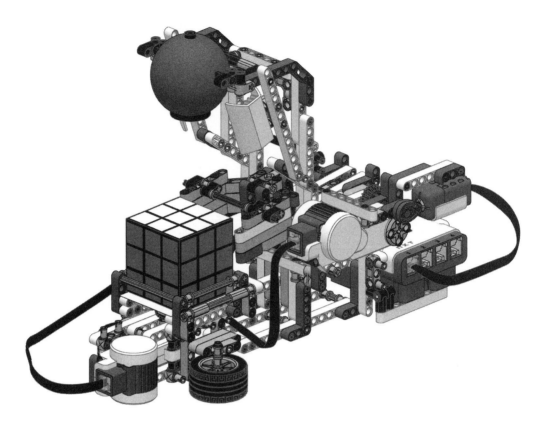

The One-Armed Wonder is ready to solve the Rubik's Cube!

thinking robots

This short appendix is designed for readers who want to deepen their understanding of artificial intelligence as it applies to problem solving and game playing. Rather than cover the topic in great detail (which is far beyond the scope of this book), I'll give you some hints and suggestions so you can continue your own research. I'll also recommend some resources that will give you an idea of how a computer can solve problems and play games successfully.

how can a machine solve problems and play games?

The robots in this book think that they are able to play games like tic-tac-toe or solve problems like a shuffled Rubik's Cube. But behind that thinking behavior are algorithms that search for the best solution to each problem.

For our purposes, the *problem* that confronts the robot is a collection of information that the machine must do something with. The basic elements of the problem are *states* and *actions*. For example, in the case of the Rubik's Cube problem, the states are all possible cube configurations, and the actions are all possible face rotations. In the case of tic-tac-toe, the states are all allowable board configurations, and the actions are the players' moves.

In order to solve a problem, the robot needs to know the problem's *initial state* (the state of the shuffled cube or that of the tic-tac-toe board after the opponent's first move) and the *set of all possible actions* that the algorithm can apply to change the problem's state (such as single face rotations of the cube and marks on the tic-tac-toe board).

The problem's *state space* is the set of all states reachable from the initial state by any sequence of actions. A *path* in the state space is any sequence of actions that leads from one state to another.

Another key element of a problem is the *goal test*, which is used to determine if a state is a *goal state*. For example, the machine reaches the goal state of tic-tac-toe when one of the players wins or the board is full (a draw game), while the goal state for solving the Rubik's Cube is a solved cube, with all faces showing the same color. For a chess playing robot, the goal state would be checkmate, when the opponent can't move the king anymore.

You could write an algorithm that explores the entire state space of a problem using brute force, basically testing every option, but such an approach would take a huge amount of time, even when solving the simplest of problems using the most powerful computers on earth. For example, consider the fact that there are 43,252,003,274,489,856,000 possible Rubik's Cube configurations, which is approximately forty-three quintillion, a number that we cannot even imagine. For the simpler tic-tac-toe problem, "only" 362,880 possible games exist!

For this reason, the algorithms used with our thinking robots need a way to choose one path over another. They do this by evaluating the state distance from the goal and the cost required to take particular actions along a certain path. It is precisely at this point that *heuristic functions* are used to allow the robot to estimate the distance between the actual state and the goal. For example, a heuristic evaluation of the distance of a Rubik's Cube state could be the number of out of place facelets, and the path cost could be the number of rotations done from the *root node* (the first node in the tree, which has no parents) so far.

The problem-solving algorithms involved perform a *heuristic search* along a search tree. A search tree is like a living tree turned upside down. The *root state* is on top, and it is the beginning state of the search. In the case of the cube, for example, the root node is the shuffled cube to be solved. The search algorithm goes down through the branches until it reaches the goal state, which is the equivalent of one of the leaves.

Different strategies exist to optimize the search in terms of time and space (specifically, computer memory). For example, the best Rubik's Cube solving algorithms (like the one by Herbert Kociemba) store the heuristic function values of the states in a database.

One particular version of a tree search algorithm is the MinMax algorithm, which is used to make machines play two-player zero-sum games. In these games, such as in tic-tac-toe, the final score can only be 1 in the case of victory, –1 in the case of defeat, and 0 in the case of a draw, but the sum of the players' scores is 0. (See Figure A-1.)

The search tree is built by putting the various game states into the nodes of the tree and the actual game state into the root. The algorithm generates all possible successive games for each node, and then chooses the move that minimizes the opponent's score and maximizes its own score. Figure A-2 shows an example of the MinMax algorithm at work.

Suppose a game being played has a maximum of only two possible moves per player per turn. In this case, the algorithm generates a tree, circles represent the moves of the AI player (*maximizing player*), and the squares represent the moves of the human opponent (*minimizing player*). For the sake of simplicity, the algorithm's maximum depth is fixed at 4.

The algorithm evaluates each *final node* (the leaves) that will assume one of the three values (1, 0, or –1, which represent victory, draw, or defeat, respectively) from the machine's point of view. The algorithm then continues to evaluate the maximum and minimum values of the *child nodes* (nodes that are below the tree) until it reaches the root node, where it chooses the move with the largest value. This is the move

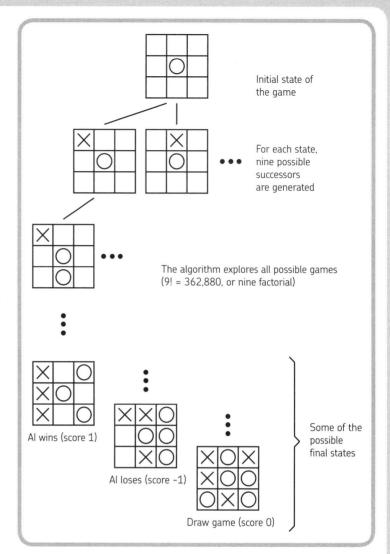

Figure A-1: The search tree for the tic-tac-toe game

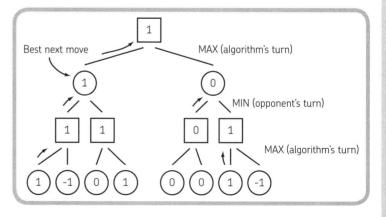

Figure A-2: A simple example of the MinMax algorithm

that the player should make in order to reduce the maximum possible loss.

We have seen only the tip of the iceberg here. However, if this short introduction to artificial intelligence has tickled your fancy, you can continue the study of the topic using the following book and web resources as your guide.

resources

Artificial intelligence is not a topic that takes ten minutes to learn. There are many branches of study, but there's one book that's recognized as the standard handbook.

Stuart J. Russell and Peter Norvig. *Artificial Intelligence: A Modern Approach.* Upper Saddle River, NJ: Prentice Hall, 2003.

Russell and Norvig's book is a great place to begin one's study of the vast field of artifical intelligence. The book presents artificial intelligence as a unified field, using the concept of an *intelligent agent* as a transversal theme. The authors cover many topics, including problem solving, machine reasoning, planning, and learning. You can see the table of contents at *http://aima.cs.berkeley.edu/*. The book is designed for advanced university students and graduate researchers.

For a gentler introduction to the world of artificial intelligence, try the Web. Here are some suggestions:

http://en.wikipedia.org/wiki/Minimax

http://en.wikipedia.org/wiki/Game_theory

http://en.wikipedia.org/wiki/Artificial_intelligence

http://en.wikipedia.org/wiki/Talk:Artificial_intelligence/ Textbook_survey

http://robotics.benedettelli.com/snail.htm

http://kociemba.org/cube.htm

credits for the CubeSolver program

The CubeSolver program is written in C++. It uses Herbert Kociemba's algorithm (*http://kociemba.org/*), as implemented by Michael Reid (*http://www.math.ucf.edu/~reid/Rubik/*). The communication with the NXT is done by using the LEGO Fantom SDK, and the video capture and rendering is based on Microsoft DirectShow.

index

LEGO MINDSTORMS NXT Thinking Robots is set in Chevin. The book was printed and bound at Malloy Incorporated in Ann Arbor, Michigan. The paper is Williamsburg VIP 70# Smooth, which is certified by the Sustainable Forestry Initiative (SFI). The book uses a RepKover binding, which allows it to lay flat when open.

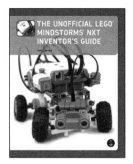

The Unofficial LEGO® MINDSTORMS® NXT Inventor's Guide

by DAVID J. PERDUE

The Unofficial LEGO MINDSTORMS NXT Inventor's Guide will teach you how to successfully plan, construct, and program robots using the MINDSTORMS NXT set, the powerful robotics kit designed by LEGO. This book begins by introducing you to the NXT set and discussing each of its elements in detail. Once you are familiar with the beams, gears, sensors, and cables that make up the NXT set, the author offers practical advice that will help you plan, design, and build robust and entertaining robots. The book goes on to cover the NXT-G programming environment, providing code examples and programming insights along the way. Rounding out the book are step-by-step instructions for building, programming, and testing six complete robots that require only the parts in the NXT set; an NXT piece library; and an NXT-G glossary.

OCTOBER 2007, 320 PP., $29.95
ISBN 978-1-59327-154-1

The LEGO® MINDSTORMS® NXT Idea Book
Design, Invent, and Build

by MARTIJN BOOGAARTS, JONATHAN A. DAUDELIN, BRIAN L. DAVIS, JIM KELLY, DAVID LEVY, LOU MORRIS, FAY RHODES, RICK RHODES, MATTHIAS PAUL SCHOLZ, CHRISTOPHER R. SMITH, *and* ROB TOROK

With chapters on programming and design, CAD-style drawings, and an abundance of screenshots, *The LEGO MINDSTORMS NXT Idea Book* makes it easy for readers to master the LEGO MINDSTORMS NXT kit and build the eight example robots. Readers learn about the NXT parts (beams, axles, gears, and so on) and how to combine them to build and program working robots like a slot machine (complete with flashing lights and a lever), a black-and-white scanner, and a robot DJ. Chapters cover using the NXT programming language (NXT-G) as well as troubleshooting software, sensors, Bluetooth, and even how to create an NXT remote control. LEGO fans of all ages will find this book an ideal jumping-off point for doing more with the NXT kit.

SEPTEMBER 2007, 368 PP., $29.95
ISBN 978-1-59327-150-3

Forbidden LEGO®
Build the Models Your Parents Warned You Against!

by ULRIK PILEGAARD *and* MIKE DOOLEY

Written by a former master LEGO designer and a former LEGO project manager, this full-color book showcases projects that break the LEGO Group's rules for building with LEGO bricks—rules against building projects that fire projectiles, require cutting or gluing bricks, or use nonstandard parts. Many of these are back-room projects that LEGO's master designers build under the LEGO radar, just to have fun. Learn how to build a catapult that shoots M&Ms, a gun that fires LEGO beams, a continuous-fire ping-pong ball launcher, and more! Tips and tricks will give you ideas for inventing your own creative model designs.

AUGUST 2007, 192 PP. *full color*, $24.95
ISBN 978-1-59327-137-4

The Unofficial LEGO® Builder's Guide

by ALLAN BEDFORD

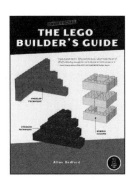

The Unofficial LEGO Builder's Guide combines techniques, principles, and reference information for building with LEGO bricks that go far beyond LEGO's official product instructions. Readers discover how to build everything from sturdy walls to a basic sphere, as well as projects including a mini space shuttle and a train station. The book also delves into advanced concepts such as scale and design. Includes essential terminology and the Brickopedia, a comprehensive guide to the different types of LEGO pieces.

SEPTEMBER 2005, 344 PP., $24.95
ISBN 978-1-59327-054-4

LEGO® MINDSTORMS® NXT One-Kit Wonders
Ten Inventions to Spark Your Imagination

by JAMES FLOYD KELLY, MATTHIAS PAUL SCHOLZ, CHRISTOPHER R. SMITH, MARTIJN BOOGAARTS, JONATHAN DAUDELIN, ERIC D. BURDO, LAURENS VALK, BLUETOOTHKIWI, *and* FAY RHODES

LEGO MINDSTORMS NXT One-Kit Wonders is packed with building and programming instructions for ten innovative robots. The book dives headfirst into the creative thrill of robot-building with models like Grabbot, Dragster, and The Hand. Step-by-step building instructions make it simple to construct even the most complex models while the detailed programming instructions teach you how a NXT program really works. And best of all, you only need one NXT Retail kit to build each of the ten robots!

NOVEMBER 2008, 408 PP., $29.95
ISBN 978-1-59327-188-6

PHONE:
800.420.7240 OR
415.863.9900
MONDAY THROUGH FRIDAY,
9 AM TO 5 PM (PST)

FAX:
415.863.9950
24 HOURS A DAY,
7 DAYS A WEEK

EMAIL:
SALES@NOSTARCH.COM

WEB:
WWW.NOSTARCH.COM

MAIL:
NO STARCH PRESS
555 DE HARO ST, SUITE 250
SAN FRANCISCO, CA 94107
USA

about the author

Daniele Benedettelli was born a quarter century ago in Grosseto, the capital of the beautiful Maremma area of Tuscany. He spent his childhood and adolescent years composing music and playing the piano. When not playing the piano, he could often be found playing with LEGO.

His passion for LEGO waned a bit when real-life interests (read "girls") got the better of him and his plastic creations. AFOLs (adult fans of LEGO) usually refer to this period as the "dark age of LEGO." But then, in 2001, he purchased a LEGO MINDSTORMS Robotics Invention System and began a career in the LEGO community.

In 2006 Benedettelli was selected by The LEGO Group as a member of the MINDSTORMS Developer Program, and he was a MINDSTORMS Community Partner from 2007 through 2009.

And in 2007, he gave birth to a LEGO NXT robot that could solve a 3 × 3 Rubik's Cube in under one minute, the result of his LEGO Rubik Utopia project.

Benedettelli holds a B.Sc. degree in Computer Engineering (specializing in Automation) and an M.Sc. degree in Robotics and Automation from the University of Siena, Tuscany. He works as a researcher in mobile robotics, and occasionally composes songs and music for spots and short movies.

software and updates

LEGO MINDSTORMS NXT Thinking Robots is supported by a companion website:

http://tr.benedettelli.com/

Here, you'll find free downloads and links to all of the programs that you'll need in order to play with your TTT Tickler and One-Armed Wonder models. The programs and links may change over time in response to reader feedback, so be sure that you're running the latest versions.

If you encounter any problems either building the robots or running the programs necessary to control them, visit *http://tr.benedettelli.com/* or *http://www.nostarch.com/nxtthinking.htm* for updates or errata. You'll find active forums for reader discussion (as well as author feedback) at *http://tr.benedettelli.com/*.

Happy building!